Beyond Ballot Boxes

Bashir Ahmed · Rafsan Jany Samir

Beyond Ballot Boxes

How Social Media Reshaped National Elections

Springer

Bashir Ahmed
Jahangirnagar University
Dhaka, Bangladesh

Rafsan Jany Samir
CloudCoder and ISTL
Dhaka, Bangladesh

ISBN 978-3-032-27349-9 ISBN 978-3-032-27350-5 (eBook)
https://doi.org/10.1007/978-3-032-27350-5

This Springer imprint is published by the registered company Springer Nature Switzerland AG
The registered company address is: Gewerbestrasse 11, 6330 Cham, Switzerland

This book is dedicated to all those who have strived to uphold democratic values and champion the power of truth in the digital age. We, Bashir Ahmed and Rafsan Jany Samir, pay tribute to the tireless efforts of our parents.
We also dedicate this work to our families and colleagues, whose unwavering support and belief in the importance of open dialogue and transparent governance have been our constant source of strength. May this book serve as a call to action for all who seek to safeguard democracy and foster an informed, inclusive public sphere in a rapidly evolving digital world.

Bashir Ahmed

Rafsan Jany Samir

Foreword

I am pleased to learn that Springer Nature is interested in publishing Prof. Bashir Ahmed and Rafsan Jany Samir's new book on *Beyond Ballot Boxes: How Social Media Reshaped National Elections.* I really appreciate their combined efforts. Given their originality, this book emerged from a deep interdisciplinary inquiry, combining insights from history, law, political science, media studies, and technology to understand how digital platforms changed, empowered, and challenged elections and socio-political systems in general.

The book is an invaluable addition to the contemporary scholarships and is genuinely thorough in its covering. The book depicts the transition from oral traditions and print media to radio, television, and finally digital platforms with artificial intelligence. This historical view lays the groundwork for understanding how media has persistently influenced public opinion, social movement, and political mobilization, despite technological and regulatory changes.

Exploring the social and cultural factors influencing media consumption in Bangladesh, the book delves into a sophisticated topic of how elements such as economy, religion, social structure, and the diaspora's influence interact with digital media, altering voter behavior and shaping political outcomes in complex and dynamic ways.

This book also takes a step forward by thoroughly exploring legislative approaches, platform accountability measures, and strategies for encouraging civil society collaborations to combat misinformation and protect electoral integrity in the digital age.

I am convinced that this book will be an effective teaching tool on media, politics, governance, and development, as well as a useful reference for future research.

Wing Hong Chui
Chair Professor of Social Work
and Criminology, Head of Department
of Applied Social Sciences
The Hong Kong Polytechnic University
Kowloon, Hong Kong

Preface

The digital revolution has dramatically transformed the way societies engage in democratic processes, reshaping the very fabric of political communication and electoral outcomes. *Beyond Ballot Boxes: How Social Media Reshaped National Elections* explores this transformation by examining the evolution of media—from the printing press to the rise of social media—and its profound impact on modern elections. This book emerged from a deep interdisciplinary inquiry, combining insights from political science, media studies, and technology to understand how digital platforms both empower and challenge democratic systems.

In Chap. 1, we introduce the central themes of the book, outlining the historical evolution of media, the rationale for studying its democratic implications, and the core research questions and objectives. The theoretical framework is established through a critical discussion of classical and contemporary models, culminating in the introduction of the Digital Democracy Disruption (D^3) Theory. Our methodology, which employs a mixed-methods approach integrating qualitative Key Informant Interviews and quantitative survey data, is also detailed.

Chapter 2 provides a comprehensive background on media evolution. It traces the journey from oral traditions and print media to the advent of radio, television, and eventually digital platforms. This historical perspective sets the stage for understanding how media has consistently influenced public opinion and political mobilization, despite shifts in technology and regulatory contexts.

In Chap. 3, the focus shifts to the challenges posed by misinformation and manipulation. We explore the global and regional dynamics of fake news, digital propaganda, and disinformation, illustrating how these issues have been amplified by technological advancements. The chapter discusses the mechanisms that enable digital propaganda, drawing comparisons between developed and developing contexts.

Chapter 4 delves into the regulatory frameworks governing digital media. By comparing the approaches of countries like the United States and those in South Asia, the chapter reveals the tension between safeguarding free speech and controlling harmful misinformation. It examines issues such as platform liability, algorithmic governance, and the delicate balance between national security and civil liberties.

Chapter 5 examines the particular case of digital media in the United States. It reviews pivotal events—from the rise of social media giants and data harvesting scandals to high-profile political campaigns—that illustrate the evolving tactics of modern electoral persuasion. This chapter provides critical insights into how data-driven strategies and micro-targeting have reshaped political campaigning.

Chapter 6 investigates the socioeconomic and cultural mediators that influence media consumption in Bangladesh. It highlights how factors such as economic status, religion, social class, and the influence of the diaspora interact with digital media, affecting voter behavior and shaping electoral outcomes in complex ways.

Chapter 7 focuses on Bangladesh's digital turn, tracing the evolution of online political engagement from 2014 to 2024. Through case studies of major elections, the chapter documents how social media has become a battleground for political mobilization, propaganda, and regulatory challenges in a transitioning democracy.

Chapter 8 offers comparative perspectives by juxtaposing the experiences of developed democracies (such as the USA and the EU) with those of developing countries (including Bangladesh, India, and Pakistan). This chapter examines structural, cultural, and technological disparities, and draws policy lessons that can help bridge these divides.

Chapter 9 proposes policy interventions aimed at strengthening democratic resilience and enhancing digital literacy. It outlines legislative approaches, platform accountability measures, and strategies for fostering civil society collaborations to combat misinformation and safeguard electoral integrity in the digital age.

Finally, Chap. 10 synthesizes the key findings and reflects on the evolving digital polity. It offers lessons learned from cross-national comparisons, provides concrete policy recommendations, and suggests future research directions for ensuring that democracy can thrive amidst rapid technological change.

Throughout this book, our analysis is both analytical and forward-looking, offering not only a detailed examination of current challenges but also actionable policy recommendations. We call on academics, policymakers, and practitioners to collaborate in building a more transparent, inclusive, and resilient democratic system that can harness the benefits of digital technology while mitigating its risks.

Dhaka, Bangladesh
March 2025

Bashir Ahmed
Rafsan Jany Samir

Competing Interests The authors have no competing interests to declare that are relevant to the content of this manuscript.

Contents

About the Authors

Bashir Ahmed is a Professor of Government and Politics at Jahangirnagar University, where he has dedicated his academic career to exploring the intersections of media, politics, and democracy. With a strong background in political communication and media studies, his research focuses on how evolving media technologies—from traditional print to modern digital platforms—shape electoral processes and influence voter behavior. Bashir has published extensively in peer-reviewed journals and edited volumes on topics ranging from digital misinformation to the transformation of public discourse in South Asia. His work not only offers historical insights but also critically examines contemporary challenges posed by digital democracy, providing valuable perspectives for policymakers and scholars alike.

Rafsan Jany Samir is a tech entrepreneur with a passion for digital innovation and its role in reshaping political landscapes. Leveraging his expertise in data analytics and digital strategy, Rafsan has been at the forefront of developing innovative solutions that bridge technology and democratic governance. His entrepreneurial ventures have focused on harnessing the power of social media to enhance voter outreach, political engagement, and transparency in elections. Through his work, Rafsan brings a pragmatic, technology-driven perspective to debates on digital democracy, advocating for ethical practices and robust regulatory frameworks to ensure that digital platforms contribute positively to democratic processes.

Chapter 1
Introduction to Digital Democracy and Social Media

1.1 Introduction

The media landscape has undergone an extraordinary evolution from the early days of print to the expansive digital networks that define the contemporary world. This book examines the profound influence of media on democratic processes and electoral outcomes, tracing its transformation from traditional platforms to modern digital ecosystems. In today's digital age, where information is disseminated at unprecedented speeds, the evolution of media is not only a historical phenomenon but also a critical factor in shaping public opinion, political engagement, and the integrity of electoral systems (Norris, 2009).

In recent years, the advent of social media and digital communication technologies has disrupted conventional media paradigms. This disruption has brought about significant changes in the way political messages are crafted, disseminated, and consumed. While these digital innovations have democratized information and empowered citizen journalism, they have also introduced new challenges most notably, the spread of misinformation, digital manipulation, and algorithmic biases (Allcott & Gentzkow, 2017). These issues, collectively conceptualized in this study as Digital Democracy Disruption (D^3), form a central focus of our inquiry. This book represents a comprehensive scholarly analysis grounded in both secondary literature and primary qualitative data. While the theoretical framework draws extensively on established research in political communication, media studies, and democratic theory, our empirical insights are derived from original Key Informant Interviews (KIIs) conducted between 2023–2025 with political analysts, journalists, civil society actors, and digital media practitioners across Bangladesh and the United States (n = 20). These interviews provide contextual depth and contemporary perspectives that complement our analysis of existing scholarship, enabling us to bridge theoretical

The original version of the chapter has been revised. A correction to this chapter can be found at https://doi.org/10.1007/978-3-032-27350-5_11

B. Ahmed and R. J. Samir, *Beyond Ballot Boxes*,
https://doi.org/10.1007/978-3-032-27350-5_1

frameworks with on-the-ground realities of digital political engagement. This mixed-methods approach, which also incorporates a targeted survey of social media users in both countries (N = 200), allows us to situate our primary data within broader scholarly conversations while offering fresh empirical insights into digital democracy's contemporary challenges (Creswell & Poth, 2018).

Studying the evolution of media and its impact on democracy is essential for several reasons. First, it allows us to understand how information flows have historically shaped public discourse and political participation. Second, examining these transformations across different contexts from developed Western democracies to emerging South Asian nations reveals both universal patterns and region-specific dynamics. Third, as we approach an era where AI-driven political propaganda, surveillance capitalism, and misinformation shape elections, understanding how media evolution impacts democracy is more critical than ever (Zuboff, 2019).

This book aims to compare media systems in the USA, UK, and South Asia, analyze the transformation of political campaigns from print to digital microtargeting, investigate government-backed surveillance and censorship, and develop a new theoretical framework (D^3) to understand digital democracy disruption. In the coming chapters, we will delve into historical media trends, contemporary election manipulation, and the geopolitical implications of digital misinformation and AI-driven politics. This is not just a theoretical discussion it is a timely analysis of how media and technology are reshaping the global democratic order.

1.2 Media as a Political Tool Across History

The 15th-century invention of the printing press revolutionized information dissemination, allowing for the mass production of newspapers, pamphlets, and books. This enabled broader political participation, fueling revolutions and shaping democratic institutions (Eisenstein, 1979). In the Western context, the American and French Revolutions were driven by print media, as revolutionary pamphlets like *Common Sense* (1776) mobilized public sentiment against authoritarian rule (Schudson, 1978). In British India, nationalist newspapers like *The Hindu* and *Amrita Bazar Patrika* played a crucial role in anti-colonial movements, highlighting print media's role in political activism. Despite its role in fostering democracy, print media is now controlled by corporate monopolies and government influence, limiting independent journalism (Ahmed, 2020).

The twentieth century saw the rise of radio and television, transforming political communication (McChesney, 1993). Radio's influence was evident through Franklin D. Roosevelt's *Fireside Chats*, which reassured American citizens during economic crises, whereas Hitler's Nazi broadcasts manipulated public sentiment (Briggs & Burke, 2009). Television introduced a new era of political spectacle, where image and media presentation became key electoral factors. The 1960 Kennedy-Nixon debate illustrated how a candidate's visual appeal could influence voter perception.

In South Asia, particularly in Bangladesh, India, and Pakistan, state-controlled broadcasters like *Doordarshan* (India) and *BTV* (Bangladesh) shaped national narratives by limiting opposition voices and controlling political discourse (Rahman, 2014).

Unlike print and electronic media, social media decentralizes political messaging, giving grassroots activists, misinformation campaigns, and authoritarian regimes equal access to global audiences (Tufekci, 2017). In Western democracies such as the USA, UK, and EU, the 2008 Obama campaign marked the first large-scale use of digital microtargeting and data-driven voter outreach (Tufekci, 2017). By 2016, the Trump campaign and Brexit referendum further demonstrated how personal data could be weaponized to manipulate voter behavior, as seen in the Cambridge Analytica scandal (Cadwalladr & Graham-Harrison, 2018). Meanwhile, in South Asian democracies like Bangladesh, India, and Pakistan, social media has become a battleground for disinformation and propaganda. The 2018 Bangladesh elections saw an overwhelming presence of bot-driven propaganda and politically motivated misinformation, influencing voter perceptions (Islam & Sultana, 2021). Similarly, in India, the spread of disinformation on WhatsApp has been strategically used to target communal tensions and influence voting behavior (Banaji & Bhat, 2020).

As we approach an era where AI-driven political propaganda, surveillance capitalism, and misinformation shape elections, understanding how media evolution impacts democracy is more critical than ever (Zuboff, 2019). This book aims to compare media systems in the USA, UK, and South Asia, analyze the transformation of political campaigns from print to digital microtargeting, investigate government-backed surveillance and censorship, and develop a new theoretical framework (D^3) to understand digital democracy disruption. In the coming chapters, we will delve into historical media trends, contemporary election manipulation, and the geopolitical implications of digital misinformation and AI-driven politics. This is not just a theoretical discussion it is a timely analysis of how media and technology are reshaping the global democratic order.

1.3 Rationale of the Study

In an era of unprecedented connectivity, the media has become both a powerful conduit for democratic discourse and a platform vulnerable to widespread manipulation. Across the world, traditional print and broadcast outlets are rapidly giving way to digital platforms where information can be created, shared, and consumed instantaneously. While this shift has brought about significant opportunities for citizen engagement, it has also introduced disruptive elements that challenge the integrity of electoral processes most notably, misinformation, algorithmic bias, and strategic political propaganda. As established in the introduction, this book seeks to uncover how these disruptions shape modern democracy, with a particular focus on South Asian contexts such as Bangladesh, India, and Pakistan, alongside comparisons to Western democracies. This chapter elaborates on the rationale behind examining media evolution and electoral impact in the twenty-first century. It also identifies key gaps in the literature, highlighting why a deeper understanding of regional specificities in South Asia is essential. By the end of this section, we establish the theoretical and practical imperative for investigating the complex relationship between media systems, voter behavior, and governance structures.

The speed at which digital media has penetrated everyday life is unparalleled in human history. Within two decades, social networking platforms and instant messaging apps have evolved from novel experiments to dominant modes of communication. Political parties, government officials, and grassroots activists now extensively utilize these tools for campaign messaging, protest organization, and policy debate. Traditional gatekeepers such as editors, publishers, and broadcasters have been partially sidelined, as user-generated content increasingly shapes public discourse. This rapid digital transformation warrants scholarly investigation, particularly in the context of emerging technologies like artificial intelligence (AI) and big data analytics, which are being deployed to micro-target voters and potentially skew electoral outcomes. Additionally, the proliferation of fake news has enabled misinformation campaigns to spread globally within seconds, eroding trust in mainstream media and democratic institutions. Moreover, algorithmic influence through automated recommendation systems on social platforms has contributed to the creation of echo chambers that reinforce partisan beliefs, fundamentally altering how voters process political information.

From the Arab Spring to the 2016 U.S. election, social media's capacity to shape political events has been well-documented. Studies such as Tufekci (2017) and Allcott and Gentzkow (2017) illustrate how platforms like Facebook and Twitter can mobilize large groups, seed protest movements, or circulate falsehoods that influence voter behavior. However, most of these seminal works emphasize Western or global contexts, leaving a gap in our understanding of developing democracies. This research is essential for balancing the geographical focus, ensuring that the unique political, economic, and cultural factors in regions like South Asia are not overlooked.

Despite the wealth of scholarship on media influence, many theoretical frameworks struggle to account for the multidimensional complexities of digital platforms. For instance, while agenda-setting theory and the public sphere concept provide valuable insights, they do not fully incorporate the rapid, decentralized flows of information now common in digital spaces. By revisiting and updating these frameworks, this study offers more nuanced analytical tools to assess digital-era electioneering and its societal impact.

A significant portion of research on media evolution and electoral impact has been conducted in Western democracies, primarily the United States and Europe. This Western-centric lens overlooks the distinct challenges faced in countries with varying levels of press freedom, literacy rates, and technological infrastructure. For example, while misinformation on Facebook or Twitter is extensively studied in the U.S., WhatsApp-driven campaigns in India or Bangladesh remain relatively underexplored, despite their massive influence in shaping voter perceptions (Banaji & Bhat, 2020).

The South Asian subcontinent encompassing diverse cultures, languages, and political systems offers a fertile ground for studying media disruption. Countries like Bangladesh are transitioning from state-controlled media to a more open environment, albeit constrained by censorship and occasional internet shutdowns. Meanwhile, India, with its enormous digital population, contends with vast misinformation campaigns that exploit religious and communal tensions. Pakistan grapples with both governmental control and militant propaganda. Yet, the academic focus on these unique dynamics remains limited, calling for in-depth, region-specific research that can illuminate how socio-political factors interplay with media evolution.

Media impact in South Asia cannot be disentangled from the region's economic, religious, and cultural underpinnings. Lower literacy levels, for instance, may heighten vulnerability to misinformation. Religious affiliations can influence how political messages are framed and received. In some cases, entire communities rely on "word-of-mouth" or localized communication apps, diverging from Western patterns of news consumption. Existing literature often lacks a deep comparative perspective that integrates these complexities with media studies and electoral politics.

South Asia is home to some of the world's largest and most complex democracies, each with its own historical trajectory of media development. Bangladesh, India, and Pakistan though politically distinct share legacies of colonialism, high population density, and socio-economic inequalities that shape media consumption. Investigating these contexts reveals how media disruptions manifest in environments where traditional hierarchies, religious affiliations, and communal tensions interplay with digital innovations.

Regulatory regimes in South Asia often oscillate between promoting media freedom and clamping down on dissent, creating volatile environments for journalists and content creators. This variability extends to internet regulations, with sporadic shutdowns, anti-fake news laws, and direct government influence on telecommunication infrastructures. Understanding how these regulations affect electoral processes is crucial for assessing the broader health of democracy in these nations.

Cultural norms heavily influence how voters engage with media. In rural areas of Bangladesh, for example, interpersonal communication can be just as influential as digital campaigns, leading to a unique blend of traditional and modern political persuasion tactics. In India, local dialects and community networks can potentiate WhatsApp-based rumor mills, shaping entire narratives around political figures. Such phenomena demand context-sensitive research that captures both digital and grassroots mobilization strategies in South Asia.

By mapping the various ways digital platforms influence elections, this study can help policymakers craft balanced regulations that uphold freedom of speech while mitigating the risks of misinformation and propaganda. For instance, governments may consider establishing independent oversight bodies to monitor election-related content on social media, or developing educational programs to boost media literacy among citizens.

Media's role in informing the public directly impacts accountability and transparency in governance. Where media systems lack independence or are overwhelmed by manipulation, citizens may lose trust in democratic institutions. A nuanced understanding of how media evolution affects these institutions is essential for strengthening electoral integrity, particularly in regions struggling with corruption, censorship, or political violence.

Educating citizens about the mechanics of misinformation, the importance of verifying news sources, and the potential biases in digital platforms is imperative. By highlighting the real-world consequences of fake news such as communal tensions, voter suppression, or policy misdirection this study underscores the urgency of public awareness initiatives. Grassroots movements, NGOs, and civil society organizations can leverage these findings to conduct awareness campaigns aimed at combating digital misinformation and fostering healthy political discourse.

The rationale for examining media evolution and its electoral impact, stressing the urgency and global relevance of the issue. It has identified key gaps in existing literature, particularly the Western bias in academic studies and the limited focus on South Asian nations. Moreover, it emphasized the role of economic, religious, and cultural factors in shaping how media phenomena manifest in different contexts.

The following chapters will build upon this rationale by outlining specific objectives and research questions, situating the study within a theoretical framework, and undertaking a detailed literature review. The regional nuances highlighted here especially the complexities of Bangladesh, India, and Pakistan will anchor the book's comparative perspective. Ultimately, by bridging these gaps and offering fresh insights into the interplay of digital media, political participation, and governance, the study aspires to contribute meaningful solutions for policy, regulatory frameworks, and public engagement in the digital age. For instance "In a world where digital media continuously disrupts established norms, understanding the interplay between media evolution and democratic processes has never been more critical. This study addresses gaps in existing literature by focusing on the distinct challenges faced by South Asian democracies, where traditional media legacies intersect with the disruptive power of digital platforms."

1.4 Objectives of the Study

1. To examine the historical evolution of media from print to electronic to digital and assess its impact on democratic processes and voter behavior across different regions.
2. To synthesize classical and contemporary theoretical frameworks (such as the public sphere, agenda-setting, and filter bubbles) to understand how media shapes public discourse and political participation.
3. To evaluate the effects of digital disruptions, including misinformation, algorithmic bias, and digital censorship, on electoral processes and democratic integrity, with a special focus on South Asia.
4. To conduct a comparative analysis of media systems and regulatory frameworks in Bangladesh, India, Pakistan, and Western democracies to identify common challenges and distinct regional dynamics.

5. To develop policy measures aimed at enhancing media literacy, improving media regulation, and safeguarding democratic processes against the adverse effects of digital media manipulation.

1.5 Key Questions

This book addresses the following key questions to guide our comprehensive analysis of digital media's impact on democracy and elections:

1. How have the historical developments in print, electronic, and digital media influenced democratic engagement and voter behavior in various political systems?
2. In what ways do theoretical frameworks such as Habermas' public sphere, agenda-setting theory, and the filter bubble concept explain the role of media in shaping public opinion and electoral outcomes?
3. In what ways has the rise of digital media transformed political participation and electoral processes in South Asian democracies, particularly in countries like Bangladesh, India, and Pakistan?
4. What are the primary challenges posed by misinformation, fake news, and digital propaganda on social media platforms, and how do these challenges affect voter behavior and political discourse in emerging democracies?
5. To what extent do economic status, religious values, literacy levels, and social class mediate the influence of digital media on voter behavior in South Asia?
6. How effective are existing media regulatory frameworks in addressing issues such as digital censorship, algorithmic bias, and governmental surveillance in both established and emerging democracies?
7. What policy interventions can be recommended to enhance media literacy, strengthen democratic resilience, and mitigate the negative effects of digital disruptions on electoral processes?

1.6 Theoretical Framework

The relationship between media and democracy has been widely explored across multiple disciplines, including political science, communication studies, sociology, and technology studies. Theories of mass communication and political engagement have long emphasized the role of media in shaping voter behavior, public discourse, and governance. Scholars have sought to understand how traditional media such as newspapers, radio, and television facilitated agenda-setting, created public spheres for debate, and influenced political mobilization (Ball-Rokeach & DeFleur, 1976; Habermas, 1989; McCombs & Shaw, 1972).

However, with the advent of digital media, social networking platforms, artificial intelligence (AI), and algorithmic curation, the scale, speed, and influence of political communication have fundamentally transformed. Traditional mass media operated within editorial constraints and journalistic ethics, but digital platforms have decentralized information flows, allowing non-traditional actors including governments, bots, trolls, and disinformation networks to amplify or distort public discourse (Benkler, 2006; Pariser, 2011). As a result, classical theories fail to fully capture the real-time disruptions and manipulations occurring in modern elections.

This segment presents a comprehensive theoretical framework that integrates classical and contemporary models to explain the evolving role of digital media in electoral processes. It begins by reviewing foundational theories such as Agenda-Setting Theory, The Public Sphere, and Media System Dependency Theory that laid the groundwork for understanding media's role in shaping voter behavior. It then introduces modern perspectives, including The Filter Bubble Hypothesis, Computational Propaganda Theory, and Algorithmic Media Influence, which explain how digital platforms and AI-driven misinformation disrupt democratic participation. Finally, this section introduces Digital Democracy Disruption (D^3) Theory, a new conceptual model that unifies these perspectives to explain algorithmic political influence, state-sponsored propaganda, and the erosion of voter autonomy in the digital age.

The Agenda-Setting Theory argues that mass media does not tell people what to think, but rather what to think about, by controlling which topics dominate public discourse (McCombs & Shaw, 1972). This theory suggests that political campaigns, government agencies, and media organizations influence elections by deciding which issues receive media coverage and how they are framed. The theory is relevant to explain the traditional media focus that has played a crucial role in shaping electoral narratives by selectively highlighting key political topics, influencing voter concerns, and directing attention toward specific candidates or policies.

On the other hand, while this theory explains traditional media influence, it does not fully capture the fragmented and decentralized nature of social media, where user-generated content, influencers, and AI-driven algorithms shape political discourse outside traditional editorial control.

Habermas (1989) introduced the Public Sphere Theory, where citizens engage in rational, democratic debate, influencing governance. In the digital age, Benkler (2006) expanded this idea into the Networked Public Sphere, where social media platforms create new spaces for political mobilization, grassroots activism, and online discourse.

This theory is essential for the Platforms like Facebook, Twitter, and YouTube facilitate public debate, political mobilization, and alternative media narratives, often bypassing state-controlled or corporate media. On the contrary, the ideal of rational discourse is often undermined by misinformation, trolling, echo chambers, and algorithmic content manipulation, which distort public debates rather than fostering democratic engagement.

The Media System Dependency Theory posits that the more people rely on media for political information, the greater its power in shaping opinions and behaviors (Ball-Rokeach & DeFleur, 1976). This is especially relevant in low-literacy regions, where voters depend heavily on television, social media, and digital news platforms.

For example, in Bangladesh, India, and Pakistan, where digital media penetration is high but media literacy remains low, many citizens rely on WhatsApp, Facebook, and YouTube as primary news sources, making them highly susceptible to misinformation (Rahman & Chowdhury, 2022).This theory does not address how AI-driven recommendation systems manipulate voter dependency, filtering what political content users see without their awareness.

The Filter Bubble Hypothesis explains how social media platforms personalize user experiences through AI-driven algorithms, reinforcing existing beliefs and preventing exposure to diverse perspectives (Pariser, 2011).Political polarization is exacerbated by filter bubbles, where voters are only shown content that aligns with their ideological preferences, creating highly segmented electorates (Allcott & Gentzkow, 2017).

Computational Propaganda Theory highlights how governments, political parties, and organizations use AI-driven bots, troll armies, and algorithmic tools to manipulate public discourse and influence elections (Woolley & Howard, 2018). For instance, the rise of bot-driven political propaganda and AI-generated disinformation plays a critical role in electoral manipulation across the USA, Europe, and South Asia (Zuboff, 2019).

After reviewing existing theories on media influence, political communication, and digital disruptions, it became evident that no single framework fully captures the complexity of contemporary electoral manipulation in the digital age. Classical theories such as Agenda-Setting, The Public Sphere, and Media System Dependency Theory provide valuable insights into media's role in shaping public discourse, but they fail to address the speed, scale, and AI-driven nature of modern disinformation, algorithmic biases, and computational propaganda. Similarly, more recent theories, including The Filter Bubble Hypothesis, Computational Propaganda Theory, and Algorithmic Media Influence, highlight critical aspects of digital disruption but do not offer a unified model that accounts for the evolving relationship between AI-driven political strategies, social media echo chambers, and state-sponsored digital interventions.

Given these limitations, this study introduces a new conceptual model Digital Democracy Disruption (D^3) to bridge these gaps. This model synthesizes and extends a body of work that has examined the intersections of technology and democratic crisis. D^3 Theory synthesizes classical and contemporary media theories while explicitly addressing the algorithmic, computational, and AI-driven disruptions reshaping democratic processes. By identifying three core disruptions algorithmic control over political narratives, the fragmentation of the public sphere into ideological silos, and the rise of AI-driven electoral manipulation D^3 provides a more comprehensive framework to analyze media's evolving role in democracy across both Western and emerging digital states.

Digital Democracy Disruption (D^3) Theory, a unified model that explains the evolving role of digital media in shaping modern elections. While traditional theories of media influence focused on agenda-setting, public discourse, and media dependency, they fail to capture the scale and speed at which digital media transforms political landscapes today. With the rise of social media, artificial intelligence (AI), and algorithmic governance, the mechanisms of political communication and electoral influence have drastically shifted. D^3 Theory provides a comprehensive framework to analyze these disruptions by identifying the structural, behavioral, and technological transformations that redefine the relationship between media and democracy in the digital age.

D^3 Theory asserts that digital media does not merely "influence" democracy it fundamentally disrupts it. Unlike previous eras, where media shaped public opinion within controlled environments, the emergence of AI-driven platforms and decentralized networks has removed traditional checks and balances, altering the way political narratives are created, disseminated, and consumed. This transformation has resulted in three primary disruptions. First, rational political discourse has been replaced by viral, emotionally charged misinformation, as political actors and interest groups leverage algorithmic amplification to spread content that maximizes engagement, often at the expense of truth and journalistic integrity. Second, there is an erosion of voter autonomy, as AI-driven behavioral micro-targeting manipulates electoral decision-making by influencing emotions, fears, and biases on an unprecedented scale. Third, digital authoritarianism is on the rise, as governments, political parties, and corporations exploit computational propaganda and digital surveillance to maintain political dominance. These disruptions collectively threaten democratic governance, electoral transparency, and informed civic participation, making it essential to reassess existing theoretical models.

One of the fundamental disruptions introduced by digital media is the Disruption of Agenda-Setting. Traditionally, agenda-setting power was held by journalists, editors, and media institutions, who determined which political issues were prioritized in public discourse (McCombs & Shaw, 1972). However, in the digital age, AI-driven platforms not human editors now decide which political topics gain visibility. Algorithms optimize content based on engagement, not credibility, which means that sensationalized, misleading, or extreme political narratives are more likely to dominate political discourse. As a result, traditional media gatekeeping has been replaced by algorithmic curation, significantly altering the formation of public opinion and electoral priorities.

Another major shift is the Disruption of the Public Sphere. The traditional public sphere a space where rational discourse and democratic deliberation take place has now been fragmented into polarized echo chambers (Benkler, 2006). Social media platforms create filter bubbles, in which individuals are primarily exposed to content that reinforces their existing beliefs and biases, preventing them from encountering alternative viewpoints (Pariser, 2011). This fragmentation is further exacerbated by coordinated disinformation campaigns, AI-generated fake news, and politically motivated bot networks, which distort reality and fuel ideological divisions. Rather than fostering meaningful democratic engagement, these digital environments increasingly serve as amplifiers of propaganda, conspiracy theories, and hyper-partisan narratives.

The third major shift is the Disruption of Political Participation, where elections are no longer shaped by policy debates and party manifestos, but by emotionally charged misinformation campaigns (Zuboff, 2019). With AI-generated deepfakes, computational propaganda, and sentiment-based micro-targeting, digital campaigns have evolved into psychological influence operations, where voters are targeted based on their fears, anxieties, and behavioral vulnerabilities rather than political ideologies or policy preferences. Political campaigns now employ advanced data analytics, machine learning models, and AI-driven advertising to influence voter sentiment in ways that were previously unimaginable. These manipulative tactics not only distort democratic decision-making but also create an unfair electoral playing field, as the most technologically advanced and well-funded actors gain disproportionate influence over voters.

D^3 Theory recognizes these three primary disruptions agenda-setting, public discourse, and political participation as the defining characteristics of digital democracy today. By analyzing how digital technologies reshape electoral influence and voter behavior, D^3 provides a unified theoretical framework that integrates insights from classical media theories, computational propaganda studies, and AI-driven political communication models. This model enables a deeper understanding of how digital disruptions are actively shaping democratic processes, both in established democracies like the United States and in emerging digital states like Bangladesh, India, and Pakistan.

Through the Digital Democracy Disruption (D^3) Theory, this study addresses these gaps by identifying three major disruptions: algorithmic control over political narratives, the fragmentation of the public sphere into echo chambers, and the rise of AI-driven electoral manipulation. These disruptions have reshaped how voters engage with political content, how misinformation spreads, and how governments and political actors exploit digital platforms to influence electoral outcomes. Unlike traditional models, which assume a structured flow of political communication, D^3 Theory acknowledges the decentralized, rapid, and unpredictable nature of digital political influence.

The implications of these disruptions are profound. AI-powered algorithms determine what political issues gain attention, often prioritizing sensationalism over factual accuracy. The public sphere is no longer a single, rational space for discourse, but a fragmented landscape of ideological silos. Political participation itself has changed, with voter decisions increasingly shaped by targeted misinformation campaigns and data-driven psychological profiling rather than informed policy debates. These developments challenge the very foundation of democratic decision-making and raise urgent questions about media regulation, platform accountability, and electoral integrity.

In the coming chapters, this book will explore how governments, political parties, and intelligence agencies leverage digital platforms to manipulate voter behavior, as well as how different regions particularly South Asia and Western democracies respond to these challenges. By critically analyzing these dynamics, this study aims to provide a comprehensive understanding of digital-era electoral manipulation and the broader consequences of media disruption in the democratic process.

1.7 Literature Review

The evolution of media has profoundly shaped democratic processes worldwide. Hallin and Mancini (2004) categorized media systems into three models: Liberal, Democratic Corporatist, and Polarized Pluralist. In the Liberal Model, exemplified by the United States and the United Kingdom, commercial interests play a dominant role in shaping news and political discourse. Media conglomerates, such as Fox News, CNN, and The New York Times, often produce sensationalized coverage, which contributes to polarized reporting and a decline in investigative journalism due to commercial pressures (McChesney, 1993). The Telecommunications Act of 1996 in the U.S. led to media consolidation, reducing media diversity and increasing profit-driven content (Bagdikian, 2004). This commercial-driven model often results in corporate agenda-setting, influencing political debates and public perception (McCombs & Shaw, 1972).

By contrast, the Democratic Corporatist Model, seen in Sweden and Norway, prioritizes public service broadcasting through outlets like Sveriges Television (SVT) and NRK. These state-supported broadcasters emphasize neutrality, public engagement, and media independence, ensuring higher levels of civic participation and trust in news media (Norris, 2009). Unlike the commercialized American model, Scandinavian nations impose strong media regulations to prevent monopolization and preserve editorial autonomy (Hallin & Mancini, 2004).

The Polarized Pluralist Model, dominant in Southern Europe and developing democracies, is characterized by media outlets closely aligned with political factions. In Italy, Spain, and Greece, political elites often influence news organizations, using them for propaganda or electoral advantage. In Italy, Silvio Berlusconi's Mediaset empire demonstrated the direct link between media ownership and political power (Hallin & Mancini, 2004). Similarly, in Spain, outlets such as El País and ABC cater to ideologically segregated audiences, reinforcing partisan divides.

Recent developments have disrupted these traditional media models due to the rise of digital platforms and social media, introducing new challenges in data privacy, misinformation, and algorithmic biases (Pariser, 2011; Sunstein, 2017). Public service broadcasters in Democratic Corporatist systems now compete with rapid online content production, complicating editorial independence (Norris & Inglehart, 2019). In Polarized Pluralist contexts, digital media allows politicians to bypass traditional news gatekeepers by directly engaging voters via Facebook, Twitter, and YouTube (Tufekci, 2017).

Beyond Europe and North America, South Asian media systems, including Bangladesh, India, and Pakistan, reveal how state control, censorship, and economic constraints further shape media landscapes (Zia, 2018). Bangladesh's transition from state-controlled television to private media channels has increased political awareness but also enabled digital censorship and intensified government surveillance (Ahmed, 2020; Rahman, 2014). Similarly, India and Pakistan face misinformation campaigns and targeted political advertising, which deepen existing social divisions (Zia, 2018).

Hallin and Mancini's (2004) comparative framework remains relevant for understanding historical and structural differences in media systems, yet it does not fully address digital disruptions, which require new analytical approaches. As digital platforms evolve, updated frameworks must account for computational propaganda, algorithmic control, and AI-driven media manipulation.

Theoretical perspectives such as Habermas' (1989) public sphere theory and McCombs and Shaw's (1972) agenda-setting theory remain foundational in explaining how media influences public opinion and electoral behavior. Habermas (1989) proposed that rational debate in the public sphere fosters democratic accountability, while McCombs and Shaw (1972) argued that media directs public attention toward specific issues although it does not dictate how people think about them.

In South Asia, public sphere theory helps explain how digital spaces (e.g., Facebook groups, WhatsApp forums) serve as modern arenas for political debate, allowing citizens to bypass state-controlled narratives. Likewise, agenda-setting theory applies to news prioritization in Bangladesh and India, where dominant TV networks influence electoral discourse by highlighting select topics (Islam & Sultana, 2021).

Critics argue that classical theories fail to capture algorithmic biases in digital communication (Pariser, 2011; Tufekci, 2017). Computational propaganda, filter bubbles, and AI-driven personalization have redefined how information is consumed, often reinforcing ideological segregation (Banaji & Bhat, 2020).

The South Asian media landscape is defined by state influence, private ownership, and digital activism (Banerjee, 2011). Countries like Bangladesh, India, and Pakistan display varying levels of press freedom and digital censorship, with social media increasingly shaping political discourse (Zia, 2018).

Bangladesh's shift from state-controlled television to privately owned media has improved political awareness but also led to increased surveillance and content regulation (Ahmed, 2020; Rahman, 2014). Facebook and YouTube are dominant political tools, used for both mobilization and misinformation (Islam & Sultana, 2021).

India's massive digital ecosystem allows for expansive political participation, but WhatsApp-driven rumor campaigns have contributed to electoral misinformation (Banaji & Bhat, 2020). During elections, mass disinformation networks amplify communal tensions and shape voter perceptions.

Pakistan faces strict media regulation, with digital activists using social platforms to challenge state narratives (Zia, 2018). However, political parties deploy troll armies to manipulate online discourse, suppress opposition, and create artificial consensus (Zia, 2018).

Comparing South Asian case studies reveals shared patterns state censorship, media bias, and social media misinformation while also highlighting unique national differences.

The rise of fake news and manipulative digital campaigns presents critical threats to global democracy (Allcott & Gentzkow, 2017). Scholars classify "information disorder" into misinformation, disinformation, and malinformation, each serving distinct political objectives (Wardle & Derakhshan, 2018).

Prolonged exposure to disinformation erodes trust in mainstream news and deepens political polarization (Sunstein, 2017). In highly divided societies, fake news influences voter behavior and even election results (Banaji & Bhat, 2020).

Fact-checking initiatives, transparency in political advertising, and media literacy programs are recommended solutions (Allcott & Gentzkow, 2017). However, regulatory interventions must balance misinformation control with freedom of speech (Wardle & Derakhshan, 2018).

Fact-checking initiatives, transparency in political advertising, and media literacy programs are recommended solutions (Allcott & Gentzkow, 2017). However, regulatory interventions must balance misinformation control with freedom of speech (Wardle & Derakhshan, 2018).

Recent scholarship emphasizes the importance of "prebunking" strategies that proactively build resistance to misinformation rather than merely correcting false claims after they spread (Lewandowsky et al., 2017).

This literature review highlights how media evolution from print to digital platforms has influenced democracy. It integrates comparative media models (Hallin & Mancini, 2004), classical theories (Habermas, 1989; McCombs & Shaw, 1972), and regional case studies, particularly in South Asia. While traditional media regulation focused on press freedom, new digital challenges, such as misinformation, AI-driven manipulation, and state surveillance, demand updated frameworks (Persily & Tucker, 2020; Suzor, 2019). The following chapters will investigate how digital democracy is simultaneously enabled and disrupted in the contemporary political landscape.

1.8 Methodology

This book is primarily a comprehensive scholarly synthesis that integrates insights from political communication, media studies, and democratic theory to examine the transformation of electoral politics in the digital age. While our analysis is grounded in extensive review of secondary literature and established theoretical frameworks, we substantially enhance this scholarly foundation with original empirical data collected specifically for this project. The Key Informant Interviews ($n = 20$) and targeted survey ($N = 200$) serve multiple important purposes: they contextualize theoretical frameworks within contemporary realities, provide expert perspectives on recent digital political developments, illustrate how abstract concepts manifest in practice, and ground our comparative analysis in voices from both Bangladesh and the United States.

This mixed-methods approach distinguishes our work from purely theoretical treatments by offering empirical grounding and contemporary relevance, while avoiding the narrow focus of single-study research reports. The empirical components complement rather than supplant the comprehensive literature synthesis that forms the backbone of each chapter. Readers will find that the book maintains analytical breadth and theoretical depth while being enriched by firsthand insights from practitioners, experts, and social media users navigating the digital political landscape.

Methodological Approach

Research Design: Mixed-Methods Approach this book integrates empirical insights with comprehensive theoretical analysis through a mixed-methods approach, combining both qualitative and quantitative techniques to examine how social media has reshaped national elections. By triangulating primary data (interviews and surveys) with secondary research (literature and document analysis), the study achieves a robust understanding of digital media's influence on political behavior (Tashakkori & Teddlie, 2010).

This convergent design allows insights from in-depth expert interviews to inform the interpretation of survey trends, and vice versa, thereby triangulating findings for greater validity (Denzin, 2012). The mixed-methods strategy is especially useful for capturing the nuanced ways in which social media impacts voter behavior across different contexts, and for comparing these dynamics between a developing democracy (Bangladesh) and a developed democracy (the USA).

Primary Data Collection

Our methodological approach integrates three distinct data sources: (1) Key Informant Interviews (KIIs) with 20 expert practitioners and observers of digital political communication in Bangladesh (n = 16) and the United States (n = 4), conducted between January 2023 and March 2025; (2) a targeted survey of social media users in both countries (N = 200: Bangladesh = 160, USA = 40) examining digital media consumption patterns and electoral engagement; and (3) systematic analysis of secondary literature spanning political science, communication studies, and technology policy domains. This triangulated approach allows us to situate our primary data within broader scholarly conversations while offering fresh empirical insights into digital democracy's contemporary challenges (Creswell & Poth, 2018; Kumar, 1989).

Key Informant Interviews

A total of 20 Key Informant Interviews (KIIs) were conducted to gain qualitative insights. Of these, 16 interviews took place in Bangladesh and 4 in the United States, ensuring input from both developing and developed democratic contexts. The interviewees were experts in political science, journalism, digital media, and civil society, selected for their knowledge of media and election dynamics. A purposive sampling strategy was used to identify key experts such as academics, senior journalists, digital campaign strategists, and NGO leaders working on governance and media literacy.

Interviews were semi-structured, following an open-ended questionnaire that allowed participants to discuss topics like the influence of social media on voter behavior, the spread of misinformation online, the effectiveness of digital campaigning, and regulatory challenges. Each interview typically lasted about 30–60 min and was conducted in the participant's preferred language (English or Bengali) via in-person meetings or video calls. With participants' consent, interviews were recorded and later transcribed for analysis. These KIIs provided rich, contextual narratives and examples of how social media is perceived to shape political engagement and public opinion in each country.

Survey

To complement the interviews, a quantitative survey was administered to measure public experiences and perceptions of digital media in politics. The survey received 200 responses in total, with a stratified emphasis on the two focus countries: 160 respondents from Bangladesh and 40 from the USA. Respondents were adult social media users, recruited through online distribution and community networks to capture a diverse range of demographics (age, gender, and education levels) in each country. The survey was delivered via a structured questionnaire (using an online form) consisting of approximately 20 questions. These questions were designed to measure digital media's influence on political behavior and covered several key areas:

Social Media Usage Patterns: e.g. primary devices for internet access, frequency of checking platforms, and the social media platform most used for political news.

Trust and Misinformation: perceptions of the reliability of political content encountered on social media, and frequency of seeing false or misleading information about candidates or issues.

Engagement and Behavior: how often respondents engage with political content (liking, commenting, sharing) and whether they verify news from social media via external sources or fact-checking.

Advertising and Algorithms: exposure to targeted political ads on social feeds and the perceived influence of these personalized ads on voting decisions or opinions.

Echo Chambers and Exposure: whether respondents feel their social media feed mainly reinforces their existing views or exposes them to diverse perspectives.

Attitudes on Regulation and Literacy: opinions on the importance of allowing all kinds of political speech on platforms (even if offensive), views on government regulation of social media content, accountability of platforms for misinformation, and the effectiveness of civil society efforts (e.g. fact-checkers) in countering online misinformation. The survey also asked about experiences with media literacy initiatives and concerns about digital surveillance or foreign influence in online political discourse. Each question was mostly close-ended (multiple choice or Likert-scale), enabling quantification of responses. The survey instrument was pilot-tested with a small group to ensure clarity of questions and then distributed in early 2025. Participants provided informed consent at the start of the questionnaire, and responses were collected anonymously. The survey data offers broad quantitative evidence of public trends in social media usage and political attitudes, which can be directly compared and contrasted with the qualitative insights from the interviews.

Secondary Sources

In addition to primary data, the research extensively utilized **secondary sources** to contextualize and support the analysis. A thorough literature review was conducted, drawing on:

Academic Books and Journal Articles: Foundational theories and prior studies on media influence, digital campaigning, misinformation, and political communication in both global and South Asian contexts.

Official Documents and Reports: These include election commission reports, legal documents (such as social media regulations and policies), and reports by international organizations or think-tanks on social media and democracy.

News Media and Press Coverage: Credible news reports and investigative articles that document real-world instances of social media impacting election narratives, cases of fake news, or government actions in Bangladesh and the USA.

Prior Cross-National Studies: Comparative studies and surveys (from sources like Pew Research or the World Bank) providing data on internet use, social media penetration, and digital political engagement in various countries. These secondary materials were collected through library research and online databases, focusing on recent publications (primarily from the last decade to capture the rise of social media) as well as historically significant works. By incorporating secondary sources, the study grounds its findings in established knowledge, identifies gaps that the current research addresses, and provides a backdrop of real-world events and statistics. For instance, secondary data on voter turnout trends, social media user demographics, and notable incidents of misinformation helped validate and interpret the primary data findings. This combination of sources ensures that the analysis is well-rounded and anchored in both current and historical understandings of media's role in elections.

Analytical Framework

The data collected were analyzed using a multi-pronged analytical framework that reflects the mixed-methods design. Qualitative and quantitative analyses were conducted separately to leverage the strengths of each approach, and then findings were integrated. Additionally, specific attention was given to social media metrics as well as cross-country comparisons during analysis. The overall goal was to extract key themes, identify statistical patterns, and then synthesize insights to answer the research questions about social media's impact on national elections.

Qualitative Analysis

All interview transcripts from the 20 KIIs were analyzed using thematic coding techniques. An inductive coding approach was applied, where transcripts were read and re-read to identify recurring ideas, concepts, and narratives emerging from the experts' discussions. Codes were assigned to segments of the text (for example, themes like "*misinformation and fake news,*" "*public political engagement* via *Facebook,*" "*algorithmic bias concerns,*" or "*regulatory challenges*" might emerge).

These codes were then grouped into broader categories and themes reflecting patterns across different interviews. The research team utilized qualitative analysis software to organize the coding (or a manual coding framework, ensuring consistency across coders). Through this process, several key themes were extracted, such as the perceived influence of social media on voter opinion formation, trust in traditional versus digital media, the role of social networks in amplifying political polarization, and the efficacy of fact-checking efforts. Once the coding was completed, the themes were examined in relation to the research questions. Thematic analysis enabled the identification of patterns and contrasts in expert opinions for instance, common concerns about online misinformation or differences in how Bangladeshi versus American experts view government regulation of social media. Representative quotes from interviewees were selected to illustrate each major theme in the finding's chapters. This qualitative analysis provided depth and context, helping explain the "why" and "how" behind trends that the quantitative data might show.

The analytical approach follows established guidelines for mixed-methods research design, ensuring methodological rigor in both data collection and interpretation (Creswell & Poth, 2018).

Quantitative Analysis

The survey data (200 responses) were analyzed using statistical methods to uncover quantitative patterns and test relationships between variables. First, descriptive statistics were calculated for all survey questions. This included frequencies and percentages for categorical responses (e.g., the percentage of respondents using Facebook versus YouTube for political news, or the proportion who encountered fake news frequently) and measures of central tendency for any scaled items (e.g., average level of trust in social media content on a 5-point scale). These descriptive results summarized the general trends in the data and allowed a straightforward comparison between Bangladeshi and U.S. respondent groups on each item. The data were then disaggregated by country to observe any notable differences; for example, what fraction of Bangladeshi respondents versus American respondents support stricter government regulation of social media, or how the frequency of social media use differs between the two groups.

Next, inferential statistical analyses were conducted to examine relationships and differences more rigorously. Depending on the nature of the variables, appropriate statistical tests were used for instance, chi-square tests for independence to see if country of respondent is significantly associated with certain responses (such as noticing targeted ads or participating in media literacy training), and t-tests or non-parametric equivalents to compare mean responses (like average concern about government monitoring) between the Bangladesh and USA samples. In addition, statistical modeling was employed on key questions: simple and multiple regression analyses were run to test hypotheses about predictors of political behavior. For example, a regression model might assess whether the frequency of social media use and trust in online content significantly predict a respondent's self-reported influence of social media on their voting decisions. These models helped isolate the impact of individual factors while controlling for others. Due to the sample size, the modeling was kept at an exploratory level, focusing on major effects rather than very fine-grained interactions. All quantitative analysis was performed using standard software (such as SPSS or R), and results were considered significant at a conventional confidence level (e.g., $p < 0.05$). The quantitative findings are presented in the results chapters with tables and charts for clarity, providing empirical evidence (percentages, correlations, significance levels) to support the discussion. This statistical analysis adds a layer of objectivity, confirming trends suggested by the qualitative data and highlighting any surprising divergences.

Social Media Metrics and Trends

As part of the analytical framework, the study also considered key social media metrics and algorithmic trends to enrich the interpretation of findings. While the research did not involve primary collection of social media platform data, it integrated insights from secondary data and the qualitative expert discussions regarding how content spreads and gains influence online. Specifically, the analysis accounted for:

Engagement Rates: Metrics such as likes, shares, comments, and overall engagement on political posts were discussed (drawing on existing reports and the interviewees' expertise) to gauge what types of political content tend to go "viral" in Bangladesh and the USA. Understanding engagement helped in interpreting survey responses about how often people interact with political content.

Algorithmic Impact: The role of platform algorithms (like Facebook's news feed algorithm or YouTube's recommendation system) was considered when analyzing perceptions of echo chambers and misinformation. Experts often noted how algorithm-driven content curation can amplify certain partisan or emotive content. These insights were used to discuss how participants' experiences (e.g., seeing mostly aligned viewpoints or frequent misinformation) might be shaped by underlying algorithms.

Viral Trends and Hashtags: The study paid attention to notable viral trends, political hashtags, or online movements during recent election cycles in each country (from 2018 2024). By referencing these examples from news reports and social media analytics studies, we contextualized the survey findings. For instance, if many respondents in Bangladesh reported encountering false information frequently, this could be linked with known misinformation waves during specific events (like an election or a political crisis reported in the news). Similarly, high engagement in the U.S. might be contextualized by referencing movements like #Election Integrity or trending election memes. By integrating these social media metrics and trend considerations, the analysis goes beyond our survey and interview data to acknowledge the digital environment in which political communication occurs. This helped ensure that interpretations of the data are grounded in how social media algorithms and user behaviors actually function, bridging the gap between numerical data and real-world social media dynamics.

Cross-Country Comparison

Given the dual-country scope of the research, a deliberate cross-country comparative analysis was woven throughout the methodology. Both data collection and analysis were structured to allow contrasting findings from Bangladesh and the United States. During data analysis, results from the two countries were systematically compared:

For qualitative data, themes from the KIIs were examined for contextual differences. For example, experts in Bangladesh often highlighted issues like internet access challenges and government monitoring, whereas U.S. experts might emphasize platform self-regulation and polarization. The analysis notes where themes converged (such as widespread concern about misinformation) and where they diverged by country.

For quantitative survey data, responses were split by country and statistical tests (as noted above) were used to identify significant differences in proportions or means. This revealed, for instance, whether Bangladeshi respondents reported higher exposure to fake news than Americans, or if Americans showed greater concern about free speech on social media than Bangladeshis. Any contrast in responses (for example, levels of trust in traditional media vs. social media) was highlighted to understand the influence of different media landscapes and democratic contexts.

The comparative lens was also applied in interpreting social media metrics and secondary information. Metrics like social media penetration rates, typical engagement levels, and regulatory environment (gleaned from secondary sources) were contrasted between the two countries to explain why certain patterns might exist. For instance, differences in literacy rates and internet infrastructure between Bangladesh and the USA are considered when comparing how social media influences rural versus urban populations. This cross-country comparison component of the methodology is crucial for addressing the research objective of understanding developing versus developed democratic contexts. By holding Bangladesh and the USA side by side, the study can better isolate which observed effects of social media on elections are universal and which are context-dependent. The comparative approach strengthens the study's conclusions by ensuring that analysis is not siloed within one country's perspective but is attentive to global versus local dynamics in digital political communication.

The comparative framework employed here draws on established traditions in comparative communication research, which emphasize the importance of contextualizing media effects within specific political and cultural environments (Brüggemann & Wessler, 2014).

Participant category	Bangladesh	USA	Total
Political analysts/scientists	5	2	7
Journalists and media professionals	4	1	5
Civil society leaders and activists	4	0	4
Digital media practitioners/campaign strategists	3	1	4
Total	16	4	20

(continued)

(continued)

Participant category	Bangladesh	USA	Total

Note Interviews were conducted between January 2023 and March 2025 via in-person meetings or video calls as part of the research project "Beyond Ballot Boxes: How Social Media has Reshaped National Elections" approved by the Biosafety, Biosecurity and Ethical Committee of Jahangirnagar University (Reference No: BBEC, JU/M 2025/ 10 (345)). Each interview lasted 30–60 min and followed a semi-structured protocol covering topics including social media's influence on voter behavior, misinformation dynamics, digital campaign strategies, and regulatory challenges. Participants were selected through purposive sampling to ensure diverse perspectives from academics, practitioners, and civil society actors with direct knowledge of digital political communication

Survey Participant Demographics (N = 200)

Demographic characteristic	Bangladesh (n = 160)	USA (n = 40)	Total (N = 200)
Age distribution			
18–25 years	56 (35%)	11 (28%)	67 (34%)
26–35 years	64 (40%)	14 (35%)	78 (39%)
36–50 years	32 (20%)	10 (25%)	42 (21%)
51+ years	8 (5%)	5 (12%)	13 (6%)
Gender			
Male	88 (55%)	19 (48%)	107 (54%)
Female	72 (45%)	21 (52%)	93 (46%)
Location			
Urban	112 (70%)	26 (65%)	138 (69%)
Rural/suburban	48 (30%)	14 (35%)	62 (31%)
Education level			
High school or less	24 (15%)	4 (10%)	28 (14%)
Some college/associate degree	48 (30%)	8 (20%)	56 (28%)
Bachelor's degree	64 (40%)	18 (45%)	82 (41%)
Graduate degree	24 (15%)	10 (25%)	34 (17%)

Note Survey was administered online in early 2025 using a structured questionnaire consisting of approximately 20 questions covering social media usage patterns, trust and misinformation perceptions, political engagement behaviors, exposure to targeted advertising, attitudes toward regulation, and media literacy experiences. Participants were adult social media users (aged 18 years and above) recruited through online distribution and community networks. The stratified sample intentionally overweighted Bangladesh (80% of respondents) to provide robust data from this understudied context while maintaining U.S. comparison capability. All participants provided informed consent, and responses were collected anonymously in accordance with the ethical clearance received from the Biosafety, Biosecurity & Ethical Committee of Jahangirnagar University

1.8.1 Research Ethics and Participant Protection

Institutional Review Board Approval

This research received formal ethical clearance from the Biosafety, Biosecurity and Ethical Committee, Faculty of Biological Sciences, Jahangirnagar University on October 26, 2025 (Reference No: BBEC, JU/M 2025/ 10 (345)). The committee reviewed and approved all research protocols for the project titled "Beyond Ballot Boxes: How Social Media has Reshaped National Elections," including interview procedures, survey instruments, informed consent processes, and data management plans. This approval confirms that the study adheres to established ethical standards for human subjects research involving voters, political activists/campaign workers, and digital media practitioners/journalists, in accordance with principles outlined in the Belmont Report (1979), institutional policies governing social science research, and international declarations on research ethics.

The approved research protocol encompasses 200–300 participants (adult human participants aged 18 years and above) across Bangladesh and the United States, focusing on individuals involved in or influenced by national elections through social media use. The committee's review confirmed this study as human social/behavioral research (non-clinical) examining experiences, behaviors, and perceptions regarding social media's impact on elections, including potential exposure to misinformation, data privacy concerns, political manipulation, and biased information through social media platforms.

Ethical Protocol Implementation

This research was designed and conducted with strict adherence to ethical principles for social science research. Several comprehensive measures were implemented to protect participants and ensure research integrity:

Informed Consent: All participants, whether interviewees or survey respondents, were provided with clear information about the purpose of the research, the nature of their participation, and their rights. Interview participants received written or verbal consent statements and explicitly agreed before proceeding. Survey respondents reviewed an introductory consent statement and indicated voluntary agreement by choosing to participate. All participants were informed that participation was entirely voluntary and that they could skip questions, pause, or withdraw at any time without consequences. The informed consent process specifically addressed potential risks including exposure to sensitive political topics and questions about misinformation experiences.

Confidentiality and Anonymization: The confidentiality of all participants was strictly maintained. For Key Informant Interviews, no personal identifying information (names, specific titles, or institutional affiliations) appears in the research outputs. Interviewees are referenced using generalized descriptors (e.g., "Senior Political Journalist, Dhaka" or "Digital Campaign Strategist, USA") or assigned anonymous codes. Direct quotes were carefully reviewed to ensure they do not inadvertently reveal participants' identities. Survey responses were collected completely anonymously with no collection of names, email addresses, IP addresses, or other personally identifiable information beyond general demographic categories (age range, education level, geographic region). All data were stored securely in password-protected, encrypted files accessible only to the research team.

Risk Assessment and Mitigation: The Biosafety, Biosecurity and Ethical Committee determined that this research involves minimal risk to participants, consistent with the category of human social/behavioral research (non-clinical). The committee's assessment acknowledged potential risks including exposure to misinformation discussions, data privacy concerns, and political manipulation topics. To mitigate these risks, the research protocol incorporated several safeguards:

(1) All interviews focused on professional expertise and observations of publicly observable political phenomena rather than sensitive personal information;
(2) No vulnerable populations were involved;
(3) The research employed no deception, manipulation, or experimental interventions;
(4) Survey questions addressed general attitudes and behaviors regarding public political communication rather than private or sensitive matters;
(5) Participants were explicitly informed of their right to decline any question or withdraw without penalty.

Data Management and Protection: Interview recordings and transcripts are stored on encrypted devices with access limited to the principal investigator (Professor Bashir Ahmed, Department of Government and Politics, Jahangirnagar University) and authorized research team members. Digital survey data is maintained on secure servers with appropriate access controls. All research materials will be retained according to institutional data retention policies and will be properly disposed of or archived following established protocols. Participants were informed of these data management practices during the consent process. The research team maintains documentation of all consent procedures, data handling protocols, and anonymization processes in compliance with the committee's requirements.

Alignment with International Standards: This ethical protocol, as approved by the Biosafety, Biosecurity & Ethical Committee of Jahangirnagar University, aligns with international standards for social science research, including:

- **The Belmont Report (1979)**—U.S. National Commission for the Protection of Human Subjects of Biomedical and Behavioral Research
- **The Declaration of Helsinki (World Medical Association,** 2013)—ethical principles for research involving human subjects
- **The Singapore Statement on Research Integrity (2010)**—global framework for responsible conduct of research

- Professional association guidelines from the American Political Science Association (APSA) and International Communication Association (ICA) regarding research ethics in political communication studies

Participant Welfare and Transparency: Throughout the research process, participant welfare remained paramount. The research protocol approved by the committee emphasized respect for persons, beneficence, and justice—the three core ethical principles of the Belmont Report. No participant reported distress or adverse effects from participation. Several participants expressed appreciation for the opportunity to contribute their expertise to research addressing important democratic challenges. The approved protocol requires that documentation of consent procedures, data management protocols, and anonymization processes be maintained and made available for review by institutional oversight bodies, journal editors, funding agencies, or other stakeholders as appropriate.

Ongoing Ethical Compliance: The research team maintains ongoing compliance with the ethical clearance requirements. Any modifications to research protocols, unexpected adverse events, or deviations from approved procedures would be promptly reported to the Biosafety, Biosecurity & Ethical Committee for review. The research was conducted between the approved timeframe, with data collection occurring from January 2023 to March 2025 under the supervision of Professor Bashir Ahmed (Contact: bashir@juniv.edu, Department of Government and Politics, Jahangirnagar University, Savar, Dhaka-1342, Bangladesh).

This comprehensive ethical framework, validated through formal review and approval by the Biosafety, Biosecurity and Ethical Committee of Jahangirnagar University, ensures that the research meets the highest standards of integrity while respecting participant autonomy, confidentiality, and well-being. The rigorous ethical protocols employed in this study, including the formal institutional review process, enhance the credibility and trustworthiness of the research findings and demonstrate adherence to both local institutional requirements and international standards for ethical research involving human participants.

1.9 Conclusion

This introductory chapter has provided a comprehensive foundation for understanding how digital media reshapes democratic processes and election outcomes. The overview of the research topic and its rationale has clarified why examining the intersections of media evolution, social contexts, and political engagement is both timely and necessary. The objectives and research questions have outlined the specific inquiries guiding this book, ensuring a focused yet holistic approach. Through the literature review, a broader scholarly context was established, highlighting historical shifts in media systems and contemporary challenges such as misinformation and algorithmic bias. Building on these insights, the theoretical framework introduced cutting-edge models particularly Digital Democracy Disruption (D^3) Theory to capture the complexities of modern-day electoral manipulation.

The D^3 framework builds upon recent scholarship examining how platform governance mechanisms shape political discourse and democratic outcomes (Suzor, 2019).

Finally, the methodology section detailed the use of mixed methods, describing how in-depth qualitative insights from Key Informant Interviews (KIIs) and quantitative data from surveys will combine with secondary research to produce a robust analysis (Creswell & Poth, 2018; Tashakkori & Teddlie, 2010).This methodological triangulation strengthens the validity of findings by allowing cross-verification of patterns across different data sources (Denzin, 2012).The comparative framework employed throughout this book draws on established traditions in cross-national communication research, which emphasize contextualizing media effects within specific political and cultural environments (Brüggemann & Wessler, 2014).

Taken together, these sections underscore the multi-dimensional nature of digital media's impact on elections. They also set the stage for exploring how technology, policy, socio-cultural norms, and economic structures intersect to influence voter behavior in both developing and developed democratic contexts.

As emphasized by recent scholarship on digital literacy and civic engagement, understanding these intersections is crucial for developing effective interventions that can strengthen democratic resilience in the face of digital disruptions (Hobbs, 2010).The case of South Asian digital politics, while understudied compared to Western contexts, offers particularly valuable insights into how digital media operates in emerging democratic systems with distinct historical, cultural, and infrastructural characteristics (Pal et al., 2016).

In the chapters that follow, we have applied this methodological and theoretical groundwork to real-world data, offering comparative perspectives, policy discussions, and implications for strengthening electoral integrity in the digital age. Chapter 2 begins this journey by tracing the historical evolution of media from pre-print oral traditions through the digital revolution, establishing the foundational context for understanding contemporary digital democracy disruptions.

Case Study

The Digital Democracy Disruption (D^3) Theory emerges from observing how democratic processes have fundamentally transformed across contexts. In 2016, the United States witnessed Cambridge Analytica's psychographic targeting reshape voter persuasion, while Bangladesh's 2018 elections saw government-organized "cyber teams" coordinate mass digital mobilization. Both instances revealed how social media platforms—originally designed for social connection—became sophisticated instruments of political influence, fundamentally disrupting traditional democratic communication channels. The D^3 Theory posits that digital media disrupts democracy through three interconnected mechanisms: algorithmic amplification of partisan content, microtargeted persuasion based on harvested data, and the erosion of traditional journalistic gatekeeping. This disruption manifests differently across contexts—in developed democracies through hyper-polarization and filter bubbles, and in developing nations through state surveillance and digital authoritarianism—yet the underlying structural transformation remains consistent.

Questions for Discussion

1. How does the Digital Democracy Disruption (D^3) Theory differ from classical theories of media effects like agenda-setting and framing?
2. What methodological challenges arise when conducting comparative research on digital media's electoral impact across vastly different political contexts?
3. How can qualitative Key Informant Interviews (KIIs) complement quantitative survey data in understanding digital political communication?
4. To what extent do Habermas's concepts of the public sphere remain relevant in the age of algorithmic curation and platform capitalism?
5. What ethical considerations should researchers prioritize when studying politically sensitive digital phenomena in authoritarian contexts?

Food for Thought

If democracy requires an informed citizenry, but algorithms prioritize engagement over accuracy, can any theoretical framework fully capture the contradictions inherent in digital political communication?

What We Learn

Theoretical frameworks must evolve to accommodate the unprecedented speed, scale, and sophistication of digital communication. The D^3 Theory offers a lens for understanding how technology fundamentally restructures power relations between citizens, states, and platforms. Rigorous methodology—triangulating qualitative and quantitative data—becomes essential for capturing the nuanced ways digital media operates across diverse cultural and political contexts. Understanding these disruptions requires interdisciplinary scholarship that bridges political science, communication studies, and technology ethics.

References

Ahmed, S. (2020). Freedom of the press in Bangladesh: Challenges and prospects. *Journal of South Asian Studies, 10*(2), 75–98.

Allcott, H., & Gentzkow, M. (2017). Social media and fake news in the 2016 election. *Journal of Economic Perspectives, 31*(2), 211–236.

Bagdikian, B. H. (2004). *The new media monopoly.* Beacon Press.

Ball-Rokeach, S. J., & DeFleur, M. L. (1976). A dependency model of mass-media effects. *Communication Research, 3*(1), 3–21.

Banaji, S., & Bhat, R. (2020). *WhatsApp vigilantes: Fake news and right-wing propaganda in India.* Oxford University Press.

Benkler, Y. (2006). *The wealth of networks: How social production transforms markets and freedom.* Yale University Press.

Briggs, A., & Burke, P. (2009). *A social history of the media: From Gutenberg to the Internet* (3rd ed.). Polity Press.

Brüggemann, M., & Wessler, H. (2014). Transnational communication as deliberation, ritual, and strategy. *Communication Theory, 24*(4), 394–414.

Cadwalladr, C., & Graham-Harrison, E. (2018, March 17). Revealed: 50 million Facebook profiles harvested for Cambridge Analytica in major data breach. *The Guardian.*

Creswell, J. W., & Poth, C. N. (2018). *Qualitative inquiry and research design: Choosing among five approaches* (4th ed.). SAGE Publications.

Denzin, N. K. (2012). Triangulation 2.0. *Journal of Mixed Methods Research, 6*(2), 80–88. https://doi.org/10.1177/1558689812437186

Eisenstein, E. (1979). *The printing press as an agent of change.* Cambridge University Press.

Habermas, J. (1989). *The structural transformation of the public sphere: An inquiry into a category of bourgeois society.* MIT Press.

Hallin, D. C., & Mancini, P. (2004). *Comparing media systems: Three models of media and politics.* Cambridge University Press.

Hobbs, R. (2010). *Digital and media literacy: A plan of action.* The Aspen Institute. https://www.aspeninstitute.org/publications/digital-media-literacy/

Islam, M., & Sultana, F. (2021). Social media, misinformation, and electoral influence in Bangladesh. *Journal of Digital Politics, 9*(3), 67–89.

Kumar, K. (1989). *Conducting key informant interviews in developing countries* (A.I.D. Program Design and Evaluation Methodology Report No. 13). U.S. Agency for International Development.

Lewandowsky, S., Ecker, U. K., & Cook, J. (2017). Beyond misinformation: Understanding and coping with the "post-truth" era. *Journal of Applied Research in Memory and Cognition, 6*(4), 353–369. https://doi.org/10.1016/j.jarmac.2017.07.008

McChesney, R. W. (1993). *Telecommunications, mass media, and democracy: The battle for the control of U.S. broadcasting, 1928–1935.* Oxford University Press.

McCombs, M. E., & Shaw, D. L. (1972). The agenda-setting function of mass media. *Public Opinion Quarterly, 36*(2), 176–187.

Norris, P. (2009). *Public sentinel: News media and governance reform.* The World Bank.

Norris, P., & Inglehart, R. (2019). *Cultural backlash: Trump, Brexit, and authoritarian populism.* Cambridge University Press.

Pal, J., Chandra, P., & Vydiswaran, V. V. (2016). Twitter and the rebranding of Narendra Modi. *Economic & Political Weekly, 51*(8), 52–60.

Pariser, E. (2011). *The filter bubble: What the internet is hiding from you.* Penguin Books.

Persily, N., & Tucker, J. A. (Eds.). (2020). *Social media and democracy: The state of the field, prospects for reform.* Cambridge University Press.

Rahman, M. (2014). *The rise of private television in Bangladesh and its impact on political awareness.* Dhaka University Press.

Rahman, S., & Chowdhury, T. (2022). Fake news and political misinformation in Bangladesh. *Journal of Media and Politics, 10*(3), 50–72.

Schudson, M. (1978). *Discovering the news: A social history of American newspapers.* Basic Books.

Sunstein, C. R. (2017). *#Republic: Divided democracy in the age of social media.* Princeton University Press.

Suzor, N. P. (2019). *Lawless: The secret rules that govern our digital lives.* Cambridge University Press.

Tashakkori, A., & Teddlie, C. (2010). *SAGE handbook of mixed methods in social & behavioral research* (2nd ed.). SAGE Publications.

Tufekci, Z. (2017). *Twitter and tear gas: The power and fragility of networked protest*. Yale University Press.

Wardle, C., & Derakhshan, H. (2018). Thinking about 'information disorder': Formats of misinformation, disinformation, and mal-information. In C. Ireton & J. Posetti (Eds.), *Journalism, 'fake news' & disinformation* (pp. 43–54). UNESCO.

World Medical Association. (2013). World Medical Association Declaration of Helsinki: Ethical principles for medical research involving human subjects. *JAMA, 310*(20), 2191–2194. https://doi.org/10.1001/jama.2013.281053

Woolley, S. C., & Howard, P. N. (Eds.). (2018). *Computational propaganda: Political parties, politicians, and political manipulation on social media*. Oxford University Press.

Zuboff, S. (2019). *The age of surveillance capitalism: The fight for a human future at the new frontier of power*. PublicAffairs.

Zia, A. (2018). Digital activism in Pakistan: Censorship and resistance. *Journal of Political Communication, 14*(2), 112–132.

Chapter 2
Background—The Evolution of Media and Its Impact on Democracy

2.1 Background the Evolution of Media and Its Impact on Democracy

Media has always served as a primary arena through which political power is articulated and contested. From the earliest forms of oral storytelling and messenger systems, through the invention of the printing press, and on to broadcast media and the digital age, each phase of media evolution has reshaped how people receive information, form opinions, and ultimately make political decisions (Hallin & Mancini, 2004; McChesney, 1993). These transformations vary significantly across global contexts: for instance, in Western democracies notably the USA media is often shaped by corporate interests and market forces, whereas in South Asia, particularly Bangladesh, it is commonly state-controlled or heavily regulated (Ahmed, 2020; Norris, 2009). In recent decades, digital platforms have radically shifted the information balance, giving governments, corporations, and local political groups new tactics to amplify or distort voter perceptions. Key changes include algorithmic curation (Pariser, 2011), state-backed disinformation (Zia, 2018), and telecom or ICT regulation as a means to curb or funnel political content (Rahman & Chowdhury, 2022). Understanding these developments is crucial for grasping how modern electoral politics operates and how democratic governance is threatened across both developed and developing nations.

Scholars have studied the evolution of media through various theoretical and empirical lenses. Some emphasize the political economy of communication, examining how ownership structures and market forces shape content (McChesney, 1993). Others focus on media systems, highlighting differences in press freedom, regulatory frameworks, and professional journalistic norms (Hallin & Mancini, 2004). In the digital era, studies increasingly scrutinize algorithmic bias, misinformation, and the intersection of social media with populist movements (Tufekci, 2017). Particularly in South Asia, research spotlights how state agencies use media regulation and surveillance to influence electoral outcomes (Ahmed, 2020). Collectively, this body

B. Ahmed and R. J. Samir, *Beyond Ballot Boxes*,
https://doi.org/10.1007/978-3-032-27350-5_2

of literature underscores that media transformations, though globally significant, occur within local, political, and cultural contexts.

2.1.1 Historical Context: Pre-print Era

Before widespread literacy and printing, societies communicated primarily through oral means, with town criers or temple proclamations carrying official decrees and myths passed down verbally (Chowdhury, 2018). In Ancient Greece, the agora facilitated direct citizen debate, but public spaces often excluded large segments of the population, such as women, slaves, and foreigners (Norris & Inglehart, 2019). In Medieval Europe, the Catholic Church dominated script production, limiting access to written knowledge (Eisenstein, 1979). Similarly, in parts of South Asia, religious and royal elites controlled manuscripts and official edicts, concentrating power in the hands of kings and clergy. The slow spread of information, combined with widespread illiteracy, allowed those with cultural or religious authority to manipulate narratives (McChesney, 1993). As a result, media gatekeeping and censorship began taking root long before modern communication infrastructures developed.

2.1.2 Print Media: The Birth of Public Opinion (15th–20th Century)

Johannes Gutenberg's mid-fifteenth century invention of the printing press revolutionized text production by enabling mass distribution of written materials and ending the monopoly held by ecclesiastical and royal elites (Eisenstein, 1979). This breakthrough spurred the rapid spread of political and religious dissent, fueling major movements like the Protestant Reformation. In the eighteenth century, revolutionary pamphlets galvanized public sentiment in North America and Europe, exemplified by Thomas Paine's Common Sense (1776), which helped inspire the American Revolution (Schudson, 1978). By the nineteenth century, penny presses in the USA broadened media access, and an increasingly literate public began exerting influence on policy debates, from labor rights to women's suffrage (Norris, 2009). Media barons later consolidated ownership, laying the groundwork for corporate control over news agendas (Hallin & Mancini, 2004).

2.1.3 Print Media in the USA and South Asia

In the USA, print media shaped national identity by providing affordable newspapers that reached urban working classes (Schudson, 1978, p. 58). Sensational "yellow journalism" emerged in the late nineteenth century, driven by fierce competition among press magnates like William Randolph Hearst and Joseph Pulitzer (McChesney, 1993, p. 24). Alongside sensationalism, investigative journalism gained prominence, exposing corruption and reinforcing the media's watchdog function (Norris, 2009, p. 32). During the twentieth century, chain ownership reduced local diversity, enabling press barons to exert significant influence on elections (McChesney, 1993, p. 28). Meanwhile, national dailies like The New York Times and The Wall Street Journal shaped broader policy discussions far beyond urban centers (Hallin & Mancini, 2004, p. 70).

In South Asia, newspapers under British colonial rule served both administrative purposes and anti-colonial activism, a duality that persisted into the post-independence era of Bangladesh (Chowdhury, 2018, p. 62). After 1971, print media in Bangladesh grew but remained vulnerable to government pressures and financial limitations (Ahmed, 2020, p. 80). Urban–rural divides in literacy hindered widespread political engagement, and successive administrations employed censorship or legal intimidation to shape coverage (Rahman, 2014, p. 45). Comparatively, while U.S. newspapers benefitted from a robust commercial market, South Asian media contended with colonial legacies, state-centric policies, and limited literacy (Norris, 2009, p. 39). Conglomeration further altered the media landscapes in both contexts, with U.S. outlets confronting digital competition and South Asian papers facing political intimidation (Ahmed, 2020, p. 84).

2.1.4 Electronic Media: The Rise of Radio and Television

Radio introduced instant communication, exemplified by Franklin D. Roosevelt's Fireside Chats and Nazi propagandist broadcasts in the 1930s (McChesney, 1993). During the Cold War, outlets like Voice of America countered Soviet-aligned messaging. Television's addition of visual elements further influenced public perception, as seen in the 1960 Kennedy-Nixon debates (McCombs & Shaw, 1972). The advent of 24-h news channels in the 1980s and 1990s, such as CNN, intensified competition for viewer attention and further shaped political narratives (Norris & Inglehart, 2019). In Western nations, commercial broadcasting thrived, while in Bangladesh, India, and Pakistan, state-owned broadcasters like BTV, Doordarshan, and PTV offered heavily regulated content favoring government positions (Rahman, 2014). Even after privatization, partisan ownership patterns often persisted (Ahmed, 2020).

2.1.5 Social Media and Democracy in the 21st Century

Platforms like Facebook, Twitter, YouTube, and WhatsApp lowered entry barriers for content creation, empowering grassroots activists but also facilitating the rapid spread of misinformation (Tufekci, 2017). In the West, the 2008 Obama campaign demonstrated how digital platforms could mobilize voters, while the Arab Spring in 2011 showcased social media's potential to catalyze protests. By 2016, social media was firmly embedded in electoral strategies, illustrated by the Trump campaign's microtargeting on Facebook and Russian-sponsored disinformation campaigns (Cadwalladr & Graham-Harrison, 2018). Tech giants responded by adding fact-checking labels or suspending contentious accounts, sparking debate over free speech versus content moderation (Benkler et al., 2020; Gillespie, 2021). Algorithmic promotion of emotional or sensational content fosters polarization, creating echo chambers that intensify partisanship (Pariser, 2011). Debates on Section 230 protections and platform responsibility persist, revealing the fragile balance between open discourse and combating hate speech or conspiracy theories (Zuboff, 2019).

2.2 Social Media in South Asia

In South Asia, surging smartphone use has transformed the political landscape by enabling mass engagement on platforms like Facebook and WhatsApp, particularly evident in India and Pakistan. In India, digital connectivity has reached unprecedented levels, with the 2019 Lok Sabha elections serving as a prime example: WhatsApp emerged as a critical tool for disseminating official campaign messages while simultaneously acting as a channel for the rapid spread of misinformation, fake news, and inflammatory rhetoric that often fueled communal tensions (Banaji & Bhat, 2020). In Pakistan, the digital realm has similarly become a battleground for political influence, where coordinated hashtag campaigns frequently reflect military or intelligence interests and are strategically used to shape national narratives, marginalize dissent, and even steer traditional media discourses (Zia, 2018). These developments in both countries underscore a broader regional trend where digital campaigns saturate social media with targeted smear tactics, factional "hashtag wars," and rapid information flows that exacerbate societal divisions and deepen polarization, thereby raising urgent regulatory and ethical questions (Reporters Without Borders, 2022). It is only later in this ongoing digital revolution that Bangladesh emerges as another critical case; during its 2018 elections, the country experienced the deployment of bot armies and the circulation of conspiratorial YouTube videos, which similarly leveraged the expansive reach of social media to manipulate public opinion and disrupt traditional political processes (Islam & Sultana, 2021).

2.3 Behind the Algorithm: How Social Media Amplifies Bias and Polarization

Over the last decade, social media platforms have often been lauded for creating more inclusive public spheres spaces where individuals from diverse backgrounds can gather, share ideas, and engage in democratic debate. However, the reality has proven more complicated. As Eli Pariser (2011) famously argued in his work on the "filter bubble," personalized recommendation systems on platforms like Facebook, YouTube, and Twitter tend to feed users content that aligns with their existing beliefs, thereby limiting exposure to alternative viewpoints. Similarly, Cass Sunstein (2017) warns that when these channels become "echo chambers," they not only concentrate like-minded individuals in digital enclaves but also escalate partisan feelings and hostility toward opposing perspectives.

Such algorithmic curation is driven by automated systems designed to maximize user engagement often by promoting sensational or polarizing material that triggers strong emotional reactions. In the context of electoral politics, this dynamic can be particularly potent. Political parties or affiliated interest groups whether in Bangladesh or elsewhere can exploit these recommendation engines to micro-target voters based on behavioral data, tailoring messages to specific demographic or psychographic profiles (Norris & Inglehart, 2019). This practice not only reinforces users' preexisting views but can also accelerate the spread of rumors and propaganda, as false or misleading content is circulated and re-circulated within ideologically homogenous networks (Allcott & Gentzkow, 2017).

The result is a fragmented civic environment where rational deliberation is frequently overshadowed by emotional appeals, misinformation, and the weaponization of social divisions. Once trapped in these filter bubbles, users are less likely to encounter fact-checks or counterarguments, fostering an environment in which "alternative facts" can thrive. Critics argue that this dynamic has led to heightened polarization, where individuals who disagree politically increasingly view one another with suspicion or animosity. In an era when political mobilization often depends on viral memes and sensationalist headlines, the depth and nuance of policy debate can be lost, replaced by soundbites optimized for maximum shares, likes, and retweets.

Moreover, the ramifications extend beyond mere political bickering. Filter bubbles can impede citizens from arriving at a shared baseline of factual information, which is critical for any functioning democracy. Public trust in media and institutions erodes as conspiracy theories or partisan spin gain traction. Civil society groups, journalists, and tech watchdogs have pushed platforms to adopt stronger measures against disinformation, such as more transparent content moderation policies, algorithmic audits, and fact-checking partnerships. However, these initiatives often run into challenges relating to platform governance, resource allocation, and accusations of bias from competing political factions.

In places like Bangladesh, where legal frameworks like the Digital Security Act already curb open discourse, the dual effect of algorithmic bias and regulatory oversight can further stifle a truly pluralistic public sphere. On one side, government-aligned actors capitalize on recommendation systems to amplify positive narratives about the administration while marginalizing critical voices. On the other side, opposition parties and activists may struggle to break through these algorithmic barriers and must instead rely on closed messaging apps or diaspora-run news portals for dissemination. Consequently, the civic space becomes less about dynamic debate and more about parallel monologues echoing within partisan silos.

Addressing algorithmic bias and the proliferation of filter bubbles, therefore, requires a multifaceted strategy. Policymakers, tech companies, and civil society must collaborate to enhance algorithmic transparency, educate the public about media literacy, and uphold rigorous journalistic standards. By introducing features such as customizable "content diversity" controls or meaningful disclaimers on targeted political ads, platforms might encourage users to venture beyond their echo chambers. At the same time, legislation that ensures accountability without stifling free expression remains a delicate balancing act. If left unchecked, these digital echo chambers threaten to deepen societal rifts and compromise the foundation of informed citizenship not just in Bangladesh, but across democracies worldwide.

Ultimately, while social media holds immense promise for inclusive engagement, it also harbors an undercurrent of deeply entrenched biases reinforced by automated systems. Recognizing and mitigating these biases is a critical step toward building a more vibrant, fact-based public discourse. Whether through better algorithmic design, improved digital literacy programs, or stronger frameworks for transparency and accountability, efforts to address filter bubbles must remain a high priority for any society that values democratic deliberation and the free exchange of ideas.

2.4 Telecom and ICT in Election Manipulation

Beyond social media, telecom infrastructures such as cell towers, ISPs, and government regulatory bodies exert significant influence (Hasan, 2023). Internet shutdowns and data throttling, especially around elections and protests, are tools used to stifle opposition. In Bangladesh, the National Telecommunication Monitoring Centre (NTMC) oversees digital traffic and collaborates with military intelligence to track so-called "anti-state" activities (Ahmed, 2020). The Bangladesh Telecommunication Regulatory Commission (BTRC) can block websites, mandate content removal, or even enact nationwide shutdowns under vague pretexts of national security (Hasan, 2023). Advanced surveillance technologies, including Pegasus spyware, allow real-time monitoring of calls, texts, and location data, undermining privacy and chilling free speech (Ahmed, 2020).

2.5 Government Censorship and "Black Laws" in Digital Media

Illiberal regimes often adopt broad fake news or cybersecurity laws to quash legitimate dissent (Voltmer, 2013). In Bangladesh, the Digital Security Act grants sweeping powers to arrest individuals for "anti-state" or "defamatory" content, prompting fear among journalists and rights activists (Reporters Without Borders, 2022). Unlike the USA or EU, where internet blackouts are rare, authorities in Bangladesh and India intermittently impose shutdowns or prosecute online speech labeled "anti-national" (Islam & Sultana, 2021). This environment restricts investigative journalism, fosters rumor-driven narratives, and heightens polarization (Sunstein, 2017). Opposition supporters often migrate to encrypted or underground platforms, further fragmenting the public sphere (Ahmed, 2020).

2.6 Conclusion

From the days of oral storytelling to the digital era's immediacy and interactivity, media evolution remains deeply intertwined with political power. Each medium print, broadcast, or digital introduces new channels for civic engagement yet also new mechanisms for manipulation. In Western democracies, corporate ownership and data-driven campaigns predominate, while in Bangladesh and other parts of South Asia, state agencies leverage censorship, surveillance, and telecom control to influence outcomes (Rahman & Chowdhury, 2022). As AI-driven propaganda and advanced micro-targeting further complicate the political landscape, maintaining a balance between media freedom and regulatory oversight becomes ever more critical. This manuscript underscores the need for vigilant scholarship that addresses how evolving media structures shape electoral integrity and democratic governance, opening new avenues of inquiry into the interplay between technology, policy, and civic life.

Case Study

The trajectory from oral traditions to social media reveals a consistent pattern: each new medium democratizes information access while simultaneously creating new forms of control. The printing press enabled Martin Luther's Reformation by breaking the Catholic Church's information monopoly, yet also facilitated mass propaganda during World War I. Radio allowed Franklin Roosevelt's fireside chats to reach millions, but also empowered fascist demagogues like Hitler and Mussolini. Television brought the Vietnam War into American living rooms, galvanizing anti-war movements, yet also enabled the highly choreographed image politics of the Reagan era.

South Asia's media evolution followed a distinct path, shaped by colonial censorship, post-independence nation-building, and recent digital leapfrogging. Bangladesh's journey from print journalism under British rule to the "Digital Bangladesh" initiative exemplifies how developing nations navigate between technological modernization and authoritarian information control.

Questions for Discussion

1. How do colonial legacies continue to shape media regulations and press freedom in South Asian democracies?
2. What parallels exist between early print censorship and contemporary social media content moderation?
3. How has the concept of "gatekeeping" transformed from traditional journalism to algorithmic curation?
4. To what extent did electronic media (radio and television) in South Asia serve state propaganda versus democratic discourse?
5. What lessons from historical media transitions can inform contemporary debates about social media regulation?

Food for Thought

Each media revolution promised democratization, yet each also enabled new forms of manipulation—is this pattern inevitable, or can society design digital systems that genuinely empower democratic participation?

What We Learn

Media evolution is not linear progress but dialectical transformation—each innovation expands communicative possibilities while introducing novel vulnerabilities. Understanding historical patterns helps contextualize contemporary challenges, revealing that debates about misinformation, censorship, and platform power echo centuries-old tensions between freedom and control. South Asia's distinct trajectory reminds us that media's democratic potential cannot be separated from broader political, economic, and cultural contexts.

References

Ahmed, S. (2020). Freedom of the press in Bangladesh: Challenges and prospects. *Journal of South Asian Studies, 10*(2), 75–98.

Ahmed, S. (2024). The 2024 Bangladesh shutdown and social media censorship. *Journal of Digital Politics, 12*(1), 32–49.

Allcott, H., & Gentzkow, M. (2017). Social media and fake news in the 2016 election. *Journal of Economic Perspectives, 31*(2), 211–236.

Banaji, S., & Bhat, R. (2020). *WhatsApp vigilantes: Fake news and right-wing propaganda in India.* Oxford University Press.

Benkler, Y., Faris, R., & Roberts, H. (2018). *Network propaganda: Manipulation, disinformation, and radicalization in American politics*. Oxford University Press.
Benkler, Y., Faris, R., & Roberts, H. (2020). *False accusation: The unfounded claim that social media censorship of conservatives has a chilling effect on free speech*. Oxford University Press.
Cadwalladr, C., & Graham-Harrison, E. (2018). *The Cambridge analytica files*. The Guardian.
Chowdhury, T. (2018). Colonial censorship and the rise of Bengali journalism. *Journal of South Asian Media Studies, 5*(2), 57–79.
Eisenstein, E. (1979). *The printing press as an agent of change*. Cambridge University Press.
Farivar, C. (2021). *Social media's role in the capitol riot: Disinformation and the path to January 6*. NPR.
Gillespie, T. (2021). *Content moderation and the hidden decisions that shape social media*. Yale University Press.
Hallin, D. C., & Mancini, P. (2004). *Comparing media systems: Three models of media and politics*. Cambridge University Press.
Hasan, R. (2023). Algorithmic bias in Bangladesh's digital media ecosystem. *Journal of Emerging Technologies, 11*(2), 39–56.
Howard, P. N. (2006). *New media campaigns and the managed citizen*. Cambridge University Press.
Howard, P. N., Ganesh, B., & Liotsiou, D. (2018). *The IRA, social media and political polarization in the United States, 2012–2018*. Oxford Internet Institute.
Isaak, J., & Hanna, M. J. (2018). User data privacy: Facebook, Cambridge analytica, and the misuse of data. *Computer, 51*(8), 56–59.
Islam, M., & Sultana, F. (2021). Social media, misinformation, and electoral influence in Bangladesh. *Journal of Digital Politics, 9*(3), 67–89.
Marwick, A., & Partin, W. C. (2021). *QAnon and digital radicalization: How conspiracy theories spread on social media*. MIT Press.
McChesney, R. W. (1993). *Telecommunications, mass media, and democracy: The battle for the control of US broadcasting, 1928 1935*. Oxford University Press.
McCombs, M. E., & Shaw, D. L. (1972). The agenda-setting function of mass media. *Public Opinion Quarterly, 36*(2), 176–187.
Norris, P. (2009). *Public sentinel: News media and governance reform*. The World Bank.
Norris, P., & Inglehart, R. (2019). *Cultural backlash: Trump, Brexit, and authoritarian populism*. Cambridge University Press.
Pariser, E. (2011). *The filter bubble: What the internet is hiding from you*. Penguin Books.
Persily, N. (2017). Can democracy survive the internet? *Journal of Democracy, 28*(2), 63–76.
Rahman, S. (2014). *The rise of private television in Bangladesh and its impact on political awareness*. Dhaka University Press.
Rahman, S., & Chowdhury, T. (2022). Fake news and political misinformation in Bangladesh. *Journal of Media and Politics, 10*(3), 50–72.
Reporters Without Borders. (2022). *World press freedom index*. Retrieved from https://rsf.org/en/index
Schudson, M. (1978). *Discovering the news: A social history of American newspapers*. Basic Books.
Sunstein, C. R. (2017). *Republic: Divided democracy in the age of social media*. Princeton University Press.
Tufekci, Z. (2017). *Twitter and tear gas: The power and fragility of networked protest*. Yale University Press.
Voltmer, K. (2013). *The media in transitional democracies*. Polity Press
Zia, A. (2018). Digital activism in Pakistan: Censorship and resistance. *Journal of Political Communication, 14*(2), 112–132
Zuboff, S. (2019). *The age of surveillance capitalism: The fight for a human future at the new frontier of power*. PublicAffairs.

Chapter 3
Misinformation and Manipulation Global and Regional Perspectives

3.1 Introduction

In the digital age, the exponential growth of social media platforms has reshaped how citizens across the world encounter and engage with political discourse. What was once the domain of professional journalists and editorial boards has become an open, participatory environment, where everyday users can create, share, and comment on content ranging from policy debates to local gossip. This democratization of information, however, has also ushered in unprecedented challenges chief among them the proliferation of misinformation, fake news, and digital propaganda.

The phenomenon of "fake news" has gained global notoriety, particularly after controversies surrounding electoral interference in high-stakes elections like the 2016 U.S. presidential race, the Brexit referendum in the United Kingdom, and repeated poll cycles in South Asia. In many developing democracies, including Bangladesh, misinformation intersects with existing social and political tensions, sometimes fueling communal violence or undermining public trust in electoral processes. Likewise, advanced democracies like the United States grapple with sophisticated disinformation campaigns orchestrated by both domestic and foreign actors, creating deep partisan divides and leaving citizens unsure about which sources of information are credible.

Critically, the proliferation of misinformation does not occur in a vacuum it directly challenges and transforms the traditional institutions that underpin democratic governance. Election commissions, courts, media regulatory bodies, fact-checking organizations, and even political parties themselves must adapt to an environment where information moves faster than verification, where viral falsehoods can shape electoral outcomes, and where institutional authority is constantly contested in digital spaces. Understanding how social media influences these institutions and how they respond is essential to comprehending the broader implications for democratic integrity.

B. Ahmed and R. J. Samir, *Beyond Ballot Boxes*,
https://doi.org/10.1007/978-3-032-27350-5_3

This chapter aims to explore the conceptual frameworks and typologies of misinformation and propaganda, as well as the mechanisms by which they spread from automated bots and troll farms to hyper-viral loops on platforms like Facebook, WhatsApp, YouTube, Twitter (X), and TikTok. We examine how democratic institutions respond to and are influenced by the spread of misinformation, analyzing the capacity of election management bodies, judicial systems, regulatory agencies, and civil society organizations to maintain their legitimacy and effectiveness in the digital age.

We also present a comparative perspective focusing on South Asia, where shared historical, linguistic, and structural contexts create both common challenges and divergent responses to digital misinformation. By examining cases from India, Pakistan, Bangladesh, Nepal, and Sri Lanka, we explore why social media's influence on democracy varies across the region despite similar developmental trajectories. The recent political upheavals in Bangladesh (July 2024) and Nepal, driven significantly by digitally mobilized youth movements, provide compelling evidence of social media's transformative power in the region.

Finally, the chapter delves into the implications of persistent misinformation how it erodes public trust in institutions, weakens social cohesion, and jeopardizes the integrity of elections. We examine the diverse set of international stakeholder efforts to counter these threats, including fact-checking initiatives, NGO campaigns, and potential regulatory measures. In doing so, we seek to offer a nuanced understanding of why misinformation thrives in the digital ecosystem, how it affects institutional functioning, and how governments, civil society, and technology platforms might collaborate to mitigate its harmful effects.

3.2 Institutional Responses to Misinformation and Social Media Influence

3.2.1 Election Management Bodies and Electoral Integrity

Election management bodies (EMBs) the institutions responsible for administering elections face unprecedented challenges in the social media age. Traditionally tasked with managing voter registration, ballot distribution, vote counting, and result announcement, EMBs now must also contend with viral misinformation about electoral processes, voter suppression tactics spread through digital channels, and coordinated campaigns to delegitimize election results before votes are even cast.

Institutional Adaptation and Capacity Constraints

In developed democracies like the United States, the Federal Election Commission (FEC) and state-level election boards have struggled to keep pace with digital campaign innovations. The decentralized nature of U.S. election administration means that responses to social media-driven misinformation vary widely by jurisdiction. Some states have established rapid-response teams to counter false claims about polling locations, voting procedures, or ballot security, while others lack the resources or technical expertise to monitor digital platforms effectively (Persily, 2017).

In South Asia, EMBs face even steeper challenges. The Election Commission of India (ECI), managing the world's largest democratic exercise, has attempted to leverage social media for voter education while simultaneously combating misinformation. During the 2019 general elections, the ECI established a Social Media Monitoring Cell and collaborated with platforms like Facebook and Twitter to remove or flag false content about Electronic Voting Machines (EVMs) and voter fraud (Chakravartty & Roy, 2015). However, the sheer volume of content in multiple languages overwhelmed verification efforts, and misinformation continued to spread rapidly through WhatsApp's encrypted channels.

Bangladesh's Election Commission has been similarly influenced by social media dynamics, though with the added complication of governmental interference. During the controversial 2018 general elections, opposition parties used social media to document alleged irregularities and voter intimidation, while pro-government accounts disseminated counter-narratives. The Commission's credibility was further undermined when internet shutdowns ostensibly for security reasons prevented real-time monitoring and reporting of electoral processes (Ahmed, 2020). The July 2024 uprising that led to the fall of the Hasina government demonstrated how social media could bypass traditional institutional channels entirely, with citizens organizing protests and documenting state violence despite official attempts at censorship.

Influence on Institutional Legitimacy

Perhaps most critically, social media has fundamentally altered how EMBs' authority is perceived and contested. In the past, an official announcement from an election commission carried significant weight; today, such announcements compete with thousands of alternative narratives circulating online. When the U.S. Cybersecurity and Infrastructure Security Agency (CISA) declared the 2020 election "the most secure in American history," the statement was immediately challenged by viral conspiracy theories that reached millions of Americans through social media, demonstrating how institutional authority can be directly undermined in digital spaces (Silverman, 2020).

This erosion of institutional trust is even more pronounced in contexts where EMBs are perceived as lacking independence. In Pakistan, the Election Commission has faced persistent accusations of bias, allegations amplified through social media during the 2018 elections when supporters of Imran Khan's PTI and opposition parties engaged in intense digital campaigns questioning the Commission's neutrality (Shahzad, 2021). Social media thus serves as both a platform for holding EMBs accountable and a mechanism for delegitimizing them through coordinated disinformation.

Judicial Institutions and Legal Frameworks

Courts and judicial systems represent another critical institutional layer influenced by social media's rise. Traditionally, courts adjudicate election disputes, interpret campaign finance laws, and protect freedom of expression. In the digital age, they must also navigate novel legal questions about online political speech, platform liability, and the boundaries between legitimate campaigning and electoral manipulation.

Judicial Responses to Digital Disinformation

In established democracies, courts have grappled with balancing First Amendment or equivalent free speech protections against the need to prevent electoral manipulation. The U.S. Supreme Court's Citizens United v. FEC (2010) decision, which expanded corporate spending on political communication, preceded the social media explosion but set a permissive framework that has complicated efforts to regulate digital political advertising. Lower courts have since handled cases involving foreign interference, bot-driven harassment, and platform content moderation policies, often struggling to apply pre-digital precedents to novel situations (Kosseff, 2019).

In India, the Supreme Court has taken a more interventionist stance, directing the ECI to increase transparency in campaign finance and regulate social media usage during elections. In 2019, the Court mandated that political parties disclose their social media advertising expenditures, a ruling intended to bring transparency to digital campaigning. However, enforcement remains weak, and many parties exploit loopholes or simply fail to comply (Banaji & Bhat, 2020).

Bangladesh's judiciary has been significantly influenced by the Digital Security Act (DSA), which criminalizes a broad range of online expression. Courts have been accused of selectively applying the law, with cases against government critics proceeding rapidly while complaints against pro-government accounts languish. This has created a chilling effect on online political discourse, as journalists, activists, and ordinary citizens self-censor for fear of legal repercussions. The Supreme Court of Bangladesh has issued some rulings protecting freedom of expression, but lower courts often defer to executive preferences in politically sensitive cases (Human Rights Watch, 2021).

Social Media's Influence on Judicial Independence

Beyond specific rulings, social media has affected judicial independence itself. Judges increasingly face organized online campaigns aimed at influencing their decisions. In Pakistan, Supreme Court judges hearing cases related to former Prime Minister Nawaz Sharif's corruption charges were subjected to coordinated social media attacks and doxxing attempts, raising questions about whether judicial impartiality can be maintained when judges become targets of viral campaigns (Zia, 2018). Similarly, in India, Supreme Court judges have been both lionized and vilified on social media depending on their rulings, potentially influencing judicial behavior and case outcomes.

Media Regulatory Bodies and Fact-Checking Organizations

Traditional media regulatory bodies agencies that oversee broadcast standards, licensing, and content regulations have found their authority circumvented by social media platforms that operate across borders and resist national regulatory frameworks. At the same time, new institutional actors like fact-checking organizations have emerged to fill verification gaps, though their effectiveness and independence vary considerably.

Regulatory Adaptation and Jurisdictional Challenges

In the United States, the Federal Communications Commission (FCC) regulates broadcast media but has no direct authority over internet platforms due to Section 230 of the Communications Decency Act, which shields platforms from liability for user-generated content. This regulatory gap means that political advertising on television faces strict disclosure requirements while similar ads on Facebook or Twitter can circulate with minimal transparency. Proposals to amend Section 230 have been politically contentious, with debates centering on whether platforms should be treated as publishers (and thus liable for content) or neutral conduits (Kosseff, 2019).

In contrast, several South Asian countries have attempted more aggressive regulatory interventions. Pakistan's media regulator, PEMRA, initially established to oversee television and radio, has expanded its mandate to cover digital content, though enforcement remains inconsistent. Bangladesh's Bangladesh Telecommunication Regulatory Commission (BTRC) has the authority to block websites and social media platforms, powers it has used during political crises. During the July 2024 protests, the government ordered telecommunications companies to shut down mobile internet services, effectively silencing social media but also demonstrating the limits of regulatory power when citizens found workarounds through VPNs and satellite connections.

The Rise of Fact-Checking as an Institution

Independent fact-checking organizations represent a relatively new institutional response to misinformation. Organizations like the International Fact-Checking Network (IFCN) have certified fact-checkers operating in dozens of countries, including South Asia. In India, outlets like Alt News, Boom Live, and India Today's Fact Check unit have gained prominence by debunking viral misinformation, particularly false claims circulating on WhatsApp and Facebook.

However, fact-checkers face significant challenges to their effectiveness and independence. First, the volume of misinformation far exceeds their capacity to verify content, meaning they can only address a fraction of false claims. Second, fact-checks often fail to reach the same audiences as the original misinformation, limiting their corrective impact. Research suggests that fact-checks may even backfire among partisan audiences, reinforcing rather than changing beliefs a phenomenon known as the "backfire effect" (Nyhan & Reifler, 2010).

Third, fact-checkers' independence is frequently questioned, particularly in politically polarized environments. In Bangladesh, fact-checking initiatives affiliated with mainstream media outlets are viewed with suspicion by opposition supporters who perceive the media as government-controlled. During the 2018 elections, rival political camps accused fact-checkers of partisan bias, undermining their credibility, regardless of the accuracy of their work (Rahman & Karim, 2020a).

Platform Partnerships and Their Limitations

Major social media platforms have partnered with fact-checking organizations to label or reduce the distribution of false content. Facebook's Third-Party Fact-Checking Program, for instance, works with IFCN-certified organizations globally. While these partnerships have led to the labeling of some misinformation, critics argue that platforms prioritize their business models over accuracy, that fact-checking is applied inconsistently, and that labels often come too late to prevent viral spread (Vaidhyanathan, 2018).

Moreover, in countries like India and Bangladesh, fact-checking efforts must navigate complex linguistic landscapes, with content circulating in numerous languages and dialects that lack robust verification infrastructure. This linguistic diversity means that misinformation in regional languages often goes unchecked, particularly on platforms like WhatsApp where end-to-end encryption prevents external monitoring.

3.2.2 Political Parties as Institutions: Adaptation and Manipulation

Political parties are fundamental democratic institutions, serving as vehicles for aggregating interests, recruiting candidates, and organizing electoral competition. Social media has transformed how parties function, enabling new forms of mobilization but also facilitating manipulation, internal fragmentation, and the rise of personality-driven politics over institutional party structures.

Institutional Adaptation Strategies

Mainstream political parties in both developed and developing democracies have invested heavily in social media infrastructure. In the United States, both Democratic and Republican parties established sophisticated digital operations by the 2012 elections, with capabilities expanding dramatically by 2016 and 2020. These operations include data analytics teams, content creation units, micro-targeting specialists, and rapid-response teams monitoring social media in real-time (Kreiss, 2016).

In India, the Bharatiya Janata Party (BJP) pioneered aggressive social media campaigning in South Asia, establishing a vast network of IT cells at national and state levels. These cells coordinate message dissemination, manage influencer relationships, and engage in both defensive and offensive social media operations. The BJP's 2014 and 2019 electoral successes have been partially attributed to this digital dominance, prompting opposition parties to develop their own capabilities, though with varying degrees of success (Banaji & Bhat, 2020).

Bangladesh's major parties the Awami League and the Bangladesh Nationalist Party (BNP) have similarly developed cyber teams, though the ruling party's access to state resources and control over telecommunications infrastructure provides significant advantages. During the 2018 elections, the Awami League's digital operation reportedly included coordinated efforts to report and remove opposition content from Facebook while amplifying pro-government messaging (Ahmed, 2020).

Social Media's Influence on Party Institutionalization

While parties have adapted to social media, the platforms have also influenced party structures and behavior. Social media rewards individual personalities over collective party identities, enabling charismatic leaders to bypass traditional party hierarchies. In the United States, Donald Trump's mastery of Twitter allowed him to dominate news cycles and communicate directly with supporters, often contradicting official party positions. This personalization of politics has weakened party institutions' ability to enforce discipline or maintain coherent policy platforms.

In Pakistan, Imran Khan's PTI leveraged social media to build a personality-driven movement that eventually captured power in 2018. Khan's direct communication style on Twitter and Facebook allowed him to cultivate an image as an outsider challenging corrupt establishment parties, even as PTI itself developed into a conventional party structure. Following his removal from power in 2022, Khan again used social media to mobilize supporters, demonstrating how platforms enable political figures to maintain influence even when out of office (Shahzad, 2021).

Cyber Manipulation and "Troll Armies"

Perhaps most troublingly, political parties have become key actors in spreading misinformation and manipulating online discourse through organized "cyber armies" or "troll farms." These operations involve networks of paid or volunteer social media users who coordinate to amplify party messaging, harass opponents, and spread disinformation about rivals.

Research by the Oxford Internet Institute has documented organized social media manipulation in over 80 countries, with political parties being primary actors in many cases (Bradshaw & Howard, 2019). In Bangladesh, both major parties maintain cyber teams that engage in mass reporting of opposition content to trigger Facebook's automated moderation systems, effectively weaponizing platform policies to silence rivals. During election periods, these teams flood social media with memes, doctored images, and false claims about opposition candidates (Ahmed, 2020).

India has witnessed similar patterns, with the BJP's IT cell accused of coordinating harassment campaigns against journalists and activists critical of the government. Investigations have revealed networks of fake accounts that amplify pro-BJP content and attack critics, blurring the lines between legitimate political communication and coordinated inauthentic behavior (Banaji & Bhat, 2020).

This institutionalization of digital manipulation poses fundamental challenges to democratic norms. When major political parties themselves become sources of misinformation and employ tactics that undermine opponents' ability to communicate freely, the institutional foundations of competitive democracy erode. Traditional notions of parties as organizations competing within shared rules become obsolete when the rules themselves are constantly violated through digital means.

3.2.3 Civil Society and Watchdog Organizations

Civil society organizations including human rights groups, media freedom advocates, transparency watchdogs, and civic education organizations serve crucial monitoring and accountability functions in democracies. Social media has both empowered and complicated their work, enabling new forms of advocacy while exposing them to digital threats and creating new challenges in documenting and combating institutional manipulation.

Empowerment Through Digital Platforms

For civil society organizations, social media has provided unprecedented opportunities to bypass traditional media gatekeepers, mobilize supporters rapidly, and document government abuses in real-time. During the Arab Spring, organizations used Twitter and Facebook to coordinate protests and share evidence of state violence with international audiences. Similar patterns have emerged in South Asia, where organizations like Human Rights Watch, Amnesty International, and domestic watchdogs use social media to publicize issues that state-controlled or intimidated media might ignore.

In Bangladesh, civil society organizations played crucial roles during the July 2024 uprising. Groups like Odhikar (human rights documentation), the Center for Governance Studies, and various student organizations used social media to coordinate protests, share evidence of police violence, and maintain communication despite internet shutdowns. The movement's success in forcing Prime Minister Hasina's resignation demonstrated how digitally coordinated civil society can challenge even well-entrenched authoritarian regimes.

Nepal's recent political transformations similarly involved civil society organizations leveraging social media to mobilize young people disillusioned with traditional parties. Organizations focused on anti-corruption, environmental protection, and youth empowerment used TikTok, Instagram, and Facebook to reach Generation Z, contributing to broader political instability and demands for systemic change (Singh, 2020).

Digital Threats and Institutional Vulnerability

However, civil society's increased digital presence has also exposed these organizations to new forms of attack. State and non-state actors use social media to delegitimize civil society organizations, spreading conspiracy theories about their funding sources, accusing them of treason or foreign manipulation, and organizing online harassment campaigns against their staff.

In India, organizations critical of the Modi government have been subjected to coordinated social media attacks questioning their legitimacy and patriotism.

Amnesty International India, before suspending operations in 2020, faced sustained online campaigns accusing it of anti-national activities, claims amplified by pro-government influencers and eventually cited by authorities in legal actions against the organization (Banaji & Bhat, 2020).

Pakistan's civil society faces similar dynamics, with human rights organizations and media freedom advocates accused on social media of being foreign agents when they criticize military or government policies. The intermingling of genuine criticism, organized disinformation, and genuine public sentiment makes it difficult for these organizations to defend their reputations and maintain public support (Zia, 2018).

Challenges in Monitoring and Documentation

While social media enables civil society to document abuses, it also complicates verification. During crises, social media floods with images, videos, and claims that require careful verification before civil society organizations can cite them credibly. The proliferation of deepfakes and cheapfakes (simply edited media) means that even visual evidence cannot be taken at face value.

Moreover, platforms' content moderation policies can inadvertently harm civil society documentation efforts. Facebook and YouTube's automated systems often remove graphic content documenting human rights abuses, claiming violations of community standards. This "over-moderation" has led to the loss of crucial evidence of war crimes, police violence, and other abuses, with civil society organizations racing to archive content before platforms remove it (Gillespie, 2018).

Institutional Coordination and Transnational Networks

One positive development has been civil society's ability to form transnational networks through social media, enabling rapid information sharing and coordinated advocacy. When Bangladesh imposed internet shutdowns during the July 2024 protests, international civil society organizations amplified domestic voices, applied diplomatic pressure, and provided technical assistance for circumventing censorship. These networks demonstrate how digital platforms can strengthen institutional capacity across borders, though they also reinforce accusations that domestic civil society is foreign-controlled.

3.2.4 *Synthesis: Institutional Transformation in the Digital Age*

The analysis above reveals that social media's influence on democratic institutions is multifaceted and profound. Rather than simply adding new communication channels to existing institutional structures, social media has fundamentally altered how these institutions function, how their authority is perceived, and their capacity to fulfill democratic roles.

Patterns Across Institutions

Several common patterns emerge across different institutional types:

Authority Erosion

Traditional institutional authority whether of election commissions announcing results, courts issuing rulings, or fact-checkers verifying claims increasingly competes with alternative narratives circulating on social media. Institutional pronouncements no longer end debates but often intensify them.

Adaptation Disparities

Institutions vary widely in their capacity to adapt to digital environments. Well-resourced organizations in developed democracies struggle; those in developing countries with limited budgets and technical expertise face even steeper challenges. This creates vulnerabilities that can be exploited by actors with resources to invest in digital manipulation.

Weaponization of Platforms

Social media platforms designed for connection and communication are routinely weaponized against institutions. Mass reporting features are used to silence opponents, algorithms are gamed to amplify propaganda, and coordinated campaigns delegitimize institutional actors, turning tools of democratic participation into instruments of institutional undermining.

Transparency Paradoxes

While social media promises transparency enabling citizens to monitor institutions and hold them accountable it simultaneously generates opacity through the sheer volume of information, the prevalence of misinformation, and the "black box" nature of algorithmic curation. Citizens may feel more informed while actually being more confused.

Context-Dependent Impacts

The degree to which social media influences institutions depends heavily on context. In countries with strong rule of law, independent media, and established democratic norms (like much of Western Europe), institutions prove more resilient. In transitional or authoritarian contexts (like Bangladesh or Pakistan), where institutions are weaker and more susceptible to political manipulation, social media's disruptive effects are more pronounced.

Implications for Democratic Governance

The transformation of democratic institutions under social media's influence carries profound implications. If citizens cannot trust election results, court rulings, or verified information, the foundations of democratic governance crumble. When political parties prioritize viral manipulation over policy deliberation, and when civil society organizations must spend resources defending against digital attacks rather than advancing their missions, democratic quality deteriorates.

Yet the picture is not entirely bleak. Institutions are adapting, developing new capacities, and learning from failures. The question is whether this adaptation can occur quickly enough and comprehensively enough to preserve democratic integrity in the face of rapidly evolving digital threats. The following sections explore how these dynamics play out in specific regional contexts, beginning with South Asia's unique challenges and recent upheavals.

3.3 Regional Case Insights: South Asia's Shared Challenges and Divergent Responses

3.3.1 Justifying the South Asian Comparison: Shared Historical, Cultural, and Structural Context

Before examining individual country cases, it is essential to justify why South Asian nations India, Pakistan, Bangladesh, Nepal, Sri Lanka, and others are grouped together for comparative analysis. While these countries exhibit significant differences in size, political systems, and economic development, they share crucial commonalities that make regional comparison analytically valuable and that help explain similar patterns in how social media influences their political systems.

Shared Colonial Legacy and Post-colonial State Formation

All South Asian nations (except Nepal, which was never formally colonized) share experiences of British colonial rule, which shaped their administrative structures, legal systems, and political cultures. Post-independence, these nations adopted similar institutional frameworks parliamentary systems (India, Bangladesh, Pakistan at times), written constitutions emphasizing fundamental rights, and nominally independent judiciaries. However, colonial legacies also included weak state capacity, ethnic and religious divisions exploited for administrative convenience, and extractive economic structures that perpetuated inequality (Voltmer, 2013).

These shared historical experiences created similar institutional vulnerabilities that social media now exploits. Laws originally designed for colonial control such as sedition provisions, colonial-era censorship regulations, and emergency powers have been adapted for digital contexts, often to restrict online expression. The continuity of colonial-era control mechanisms into the digital age represents a distinctly South Asian pattern.

Religious and Ethnic Diversity Creating Communal Tensions

South Asia is characterized by extraordinary religious and ethnic diversity, with major populations of Hindus, Muslims, Buddhists, Christians, and Sikhs, alongside countless linguistic and ethnic groups. This diversity, combined with histories of partition (India-Pakistan in 1947, Bangladesh-Pakistan in 1971) and ongoing territorial disputes (Kashmir, Tamil regions), creates persistent communal tensions that social media amplifies (Rahman & Karim, 2020b).

Misinformation in South Asia frequently exploits these fault lines, with false claims about religious desecration, minority persecution, or foreign conspiracies spreading rapidly through WhatsApp groups organized along communal lines. The 2018 mob lynchings in India, triggered by false WhatsApp rumors about child kidnappers, demonstrated how digital misinformation could translate into violence in contexts of existing communal anxiety. Similar patterns appear across the region, justifying their comparative examination.

Rapid Digital Expansion with Low Digital Literacy

South Asian countries have experienced explosive growth in internet access over the past decade, driven by affordable smartphones and competitive telecommunications markets. India alone added hundreds of millions of internet users between 2015 and 2020, many in rural areas accessing the internet for the first time through mobile devices. Bangladesh, Pakistan, and Sri Lanka have witnessed similar trajectories, with internet penetration growing from single digits to significant portions of their populations (Rahman & Tania, 2020a).

Critically, this digital expansion has outpaced digital literacy development. Millions of first-time internet users lack the critical thinking skills to evaluate online information, having no prior experience with digital media and often limited formal education. This creates massive populations vulnerable to misinformation, a challenge shared across South Asia but largely absent in developed democracies where internet adoption occurred more gradually among more educated populations.

WhatsApp Dominance and Encrypted Messaging Challenges

Unlike Western democracies where Facebook and Twitter dominate political discourse, South Asia is characterized by WhatsApp's overwhelming dominance for political communication. WhatsApp's end-to-end encryption, group chat features, and integration with mobile phones make it ideal for organizing and spreading information but also make it nearly impossible for authorities, platforms, or fact-checkers to monitor and counter misinformation (Banaji & Bhat, 2020).

This creates a distinctly South Asian challenge: viral misinformation spreading through private WhatsApp groups that fact-checkers cannot access, regulators cannot monitor, and platforms claim they cannot moderate without compromising encryption. The dynamics of misinformation thus differ fundamentally from the public timeline-based systems dominant in Western contexts.

Weak Regulatory Capacity and Politicized Enforcement

South Asian governments generally lack the technical capacity and resources to effectively regulate social media, even when political will exists. Regulatory agencies are understaffed, lack digital expertise, and often operate with outdated mandates designed for broadcast media. When regulations do exist like Bangladesh's Digital Security Act or Pakistan's Prevention of Electronic Crimes Act enforcement is often politicized, targeting government critics while ignoring similar violations by supporters (Human Rights Watch, 2021; Zia, 2018).

This pattern of weak but politicized regulation distinguishes South Asia from both the largely unregulated U.S. model and the more systematic European regulatory approaches. It creates environments where social media operates with few meaningful constraints except when governments invoke draconian measures like internet shutdowns.

Youth Demographics and Political Transformation

South Asian countries are characterized by young populations, with median ages in the mid-to-late twenties. This demographic reality means that large segments of voters have known only the digital age, communicating primarily through smartphones and social media. Traditional media newspapers and television hold less sway among these younger voters, who receive political information through Instagram, TikTok, and WhatsApp.

This generational shift has enabled rapid political transformations, as seen in Bangladesh's July 2024 uprising and Nepal's recent instability, where Gen Z activists used social media to organize movements that toppled or challenged entrenched political orders. The youth demographic thus represents both a commonality across South Asia and a driver of convergent patterns in how social media influences politics.

Understanding Differential Impact: Why Social Media Influence Varies Despite Commonalities

While the above factors justify grouping South Asian countries for comparative analysis, social media's influence on their political systems nevertheless varies significantly. Understanding this variation requires examining factors that differ across countries:

State Capacity and Repressive Apparatus

Countries with stronger state capacity and more developed security apparatus like India and increasingly Bangladesh (until July 2024) have greater ability to monitor, manipulate, and when desired, shut down social media. Pakistan's military intelligence has sophisticated digital capabilities, while Nepal and Sri Lanka have less developed surveillance infrastructure. This affects both how freely citizens can use social media for political organizing and how effectively states can spread propaganda or suppress dissent.

Media Ecosystem Maturity

India has a relatively developed private media sector with dozens of television news channels and digital outlets, creating a more competitive information environment even as government pressure increases. Bangladesh's media is more consolidated and government-aligned, while Pakistan's media operates under military influence. These differences in media ecosystem maturity affect whether social media serves as an alternative to mainstream media or simply amplifies existing information monopolies.

Internet Penetration and Digital Divides

Despite rapid expansion, internet penetration varies considerably across South Asia. Urban–rural divides, gender gaps in internet access, and linguistic diversity create different levels of digital political participation. India's relatively higher internet penetration means social media reaches more voters directly, while in less connected countries like Bangladesh and Pakistan, information from social media often filters through intermediaries, changing its political impact.

Political System Characteristics

India's federal structure, vibrant multi-party competition, and relatively strong institutions create different dynamics than Bangladesh's de facto single-party dominance (until recently) or Pakistan's civil-military hybrid regime. Social media's role in electoral competition, government accountability, and political mobilization thus varies based on whether it operates within competitive democracy, electoral autocracy, or military-influenced systems.

Historical Context of Political Violence

Countries with more recent and severe experiences of political violence like Sri Lanka's civil war or Pakistan's terrorism challenges have populations more sensitized to security concerns, making them more receptive to government justifications for social media restrictions. Meanwhile, countries with more stable recent histories may have populations more resistant to such controls.

These variations explain why, despite shared structural factors, social media's specific impacts differ across South Asia. India's WhatsApp lynchings, Pakistan's military-managed digital propaganda, Bangladesh's youth-led digital uprising, and Nepal's Gen Z political transformation reflect both regional commonalities and country-specific configurations of state power, civil society strength, and democratic institutions.

3.3.2 India: WhatsApp Lynchings and Electoral Propaganda

India represents perhaps the most complex case of social media's impact on democracy in South Asia, combining the world's largest democratic electorate with explosive internet growth, deep communal divisions, and increasingly authoritarian tendencies under the Modi government.

WhatsApp and Communal Violence

India gained international attention in 2018 when a series of mob lynchings were linked to false rumors spreading through WhatsApp. In several incidents, forwarded messages falsely accused individuals often Muslims or outsiders of being child kidnappers or organ traffickers, leading to vigilante violence that killed dozens. These incidents demonstrated how misinformation could translate directly into deadly violence in contexts of existing communal tensions and limited digital literacy (Banaji & Bhat, 2020).

The lynchings forced WhatsApp to implement restrictions on message forwarding in India, limiting how many times a message could be forwarded and how many groups it could be sent to simultaneously. However, these technical fixes addressed symptoms rather than causes. The underlying dynamics communal anxiety, rapid digital expansion among low-literacy populations, and weak law enforcement remained unchanged. Subsequent incidents, though perhaps less numerous, have continued to occur.

Electoral Propaganda and the BJP's Digital Dominance

India's political landscape has been transformed by the Bharatiya Janata Party's (BJP) sophisticated social media operations. Drawing lessons from the Obama and Trump campaigns in the U.S., the BJP invested heavily in digital infrastructure, creating networks of IT cells at national, state, and local levels. These cells coordinate daily messaging, produce visual content, manage influencer relationships, and engage in both positive promotion and negative attacks on opponents.

During the 2014 and 2019 general elections, the BJP's digital dominance was evident. Pro-BJP content saturated Facebook, Twitter, and WhatsApp, often featuring false or misleading claims about opposition parties particularly the Indian National Congress and its leaders. Investigation by journalists and researchers revealed coordinated networks of Facebook pages and WhatsApp groups systematically spreading pro-BJP propaganda, with content often originating from the party's IT cell before being amplified through seemingly independent accounts (Chakravartty & Roy, 2015).

Opposition parties have attempted to develop comparable capabilities, but with limited success. The Congress party's digital operations, while growing, have struggled to match the BJP's scale and sophistication. Regional parties have had mixed results, with some (like the Aam Aadmi Party) achieving notable digital mobilization but others remaining largely absent from digital spaces.

Institutional Responses and Their Limitations

The Election Commission of India (ECI) has attempted to regulate social media during election periods, establishing a Social Media Monitoring Cell and partnering with platforms to remove violative content. However, the scale of the challenge overwhelms institutional capacity. With content in dozens of languages circulating on multiple platforms especially encrypted WhatsApp the ECI can only address a fraction of problematic material.

Moreover, platform enforcement has been inconsistent and politically contentious. When Facebook removed accounts and pages in 2019 for "coordinated inauthentic behavior," the affected networks included both pro-BJP and pro-Congress accounts, but critics argued that enforcement was unequal. The platforms' dependence on the Indian market particularly Facebook's massive user base has also raised questions about whether commercial interests affect content moderation decisions.

Differential Impact Across India

Social media's influence on Indian politics varies significantly across regions, urban–rural divides, and linguistic communities. In urban areas with higher education levels, social media has enabled new forms of political engagement and accountability journalism. Outlets like Alt News and Boom Live have built followings by fact-checking viral claims, though their reach remains limited compared to the misinformation they counter.

In rural areas and among less educated populations, social media serves more as an amplifier of existing communal and political divisions. False claims spread more readily, and fact-checking is less effective. WhatsApp's dominance in these contexts as opposed to Facebook or Twitter means that misinformation circulates in more closed, trusted networks, making it both more believable and harder to counter.

3.3.3 Pakistan: Narratives of Nationalism and Military Influence

Pakistan's social media landscape is shaped by the military's dominant political role, periodic transitions between civilian and military rule, and ongoing territorial disputes with India that generate nationalist fervor. Social media in Pakistan serves simultaneously as a space for civil society activism and as a tool for military-managed propaganda.

Military Influence and Digital Propaganda

Pakistan's military particularly its Inter-Services Intelligence (ISI) and Inter-Services Public Relations (ISPR) agencies has developed sophisticated social media capabilities. These capabilities are used to shape narratives around national security, foreign policy (especially regarding India and Afghanistan), and domestic politics. Pro-military accounts on Twitter, Facebook, and YouTube produce content glorifying the armed forces, justifying military interventions in politics, and attacking critics as traitors or foreign agents (Zia, 2018).

During periods of civil-military tension such as when civilian Prime Ministers have clashed with military leadership social media campaigns have been mobilized to undermine civilian authority. Before Prime Minister Nawaz Sharif's ouster in 2017, coordinated social media campaigns highlighted corruption allegations, presented military leaders favorably in comparison, and attacked Sharif's supporters. Similar patterns preceded Imran Khan's rise to power in 2018, with social media amplifying anti-establishment sentiment that the military reportedly channeled toward installing Khan as Prime Minister (Shahzad, 2021).

Imran Khan's PTI and Digital Mobilization

Imran Khan's Pakistan Tehreek-e-Insaf (PTI) pioneered civilian social media campaigning in Pakistan, building on Khan's celebrity status and anti-corruption message to mobilize urban youth. Khan's direct communication style on Twitter resonated with young, digitally connected Pakistanis disillusioned with traditional parties. The PTI's 2013 campaign demonstrated social media's mobilization potential, even though the party fell short of power. By 2018, with both military support and enhanced digital operations, Khan successfully captured the Prime Minister's office (Shahzad, 2021).

However, Khan's government (2018–2022) faced accusations of using state resources to manipulate social media, including employing "cyber teams" to attack critics and promote government narratives. After Khan's removal in 2022, he again turned to social media to mobilize supporters, organizing protests and maintaining political relevance through digital platforms even while out of office. This demonstrates how social media enables sustained political mobilization beyond traditional institutional channels.

Blasphemy, Censorship, and Digital Repression

Pakistan's digital sphere is heavily influenced by blasphemy laws and restrictions on content deemed offensive to Islam or national security. Social media users face arrest for posts critical of the military, state institutions, or religious sensitivities. Bloggers and activists have been disappeared, arrested, or charged with blasphemy based on social media posts, creating a climate of fear that limits free expression (Zia, 2018).

At the same time, social media has been used to incite blasphemy accusations themselves, with viral campaigns targeting religious minorities or liberal Muslims. False accusations of blasphemy spreading through social media have led to mob violence, demonstrating similar dynamics to India's WhatsApp lynchings but with religious rather than communal triggers.

The Pakistan Telecommunication Authority (PTA) regularly issues internet shutdowns and blocks social media platforms during politically sensitive periods. These shutdowns, often justified on security grounds, effectively silence dissent and prevent digital organization of protests or documentation of state violence. During the 2024 protests following Khan's arrest, mobile internet was repeatedly shut down in major cities, demonstrating the state's willingness to disrupt digital communications when politically threatened.

Differential Impact: Urban versus Rural, Punjabi versus Other Ethnicities

Social media's political influence in Pakistan is heavily concentrated in urban areas, particularly Lahore, Karachi, and Islamabad, where internet penetration and education levels are higher. Rural areas, especially in Balochistan and Khyber Pakhtunkhwa provinces, have limited connectivity and lower digital literacy, meaning traditional patronage networks and local leaders remain more influential than social media.

Additionally, language matters significantly. Most political social media content in Pakistan is in Urdu or English, excluding large populations that speak Punjabi, Sindhi, Pashto, or Balochi primarily. This linguistic divide means social media's political impact varies by ethnicity and region, with urban Punjabi-speakers most exposed to digital political content.

3.3.4 *Bangladesh: Digital Mobilization and the July 2024 Uprising*

Bangladesh exemplifies both social media's potential for democratic mobilization and authoritarian governments' increasing sophistication in controlling digital spaces. The country's July 2024 uprising, which forced Prime Minister Sheikh Hasina's resignation after 15 years in power, represents a watershed moment demonstrating how youth-led, digitally coordinated movements can overcome even well-entrenched authoritarian regimes. The Path to July 2024: Escalating Digital Repression: Throughout the late 2010s and early 2020s, the Hasina government increasingly used the Digital Security Act (DSA) to suppress online dissent. Journalists, activists, students, and ordinary citizens faced arrest for Facebook posts critical of the government. Between 2018 and 2024, hundreds were charged under the DSA, creating a climate of self-censorship. Mainstream media, already heavily controlled through government pressure and patronage, largely avoided criticism of the regime (Human Rights Watch, 2021). However, suppression paradoxically strengthened digital organizing by zorcing activists to develop more sophisticated operational security. Encrypted messaging, VPN usage, and anonymous accounts proliferated. Youth activists particularly university students created networks across platforms (Facebook, Instagram, TikTok, Twitter X) that remained largely underground until activated during crises. The Awami League government also developed extensive cyber capabilities, including "cyber teams" that reported opposition content, spread pro-government propaganda, and harassed critics. During the controversial 2018 general elections, these teams worked to dominate social media narratives while opposition parties struggled to gain online traction amid selective enforcement of the DSA (Ahmed, 2020). The July 2024 Uprising: Digital Mobilization Against Authoritarianism: The immediate trigger for the July 2024 protests was a government decision to restore a controversial quota system for government jobs, which students argued was discriminatory and reinforced patronage. However, underlying grievances extended far beyond this issue: economic stagnation, unemployment, corruption, and 15 years ofzncreasingly authoritarian rule had created widespread frustration, particularly among young people. University students began organizing through social media, using Facebook events to coordinate protests, Instagram and TikTok to share visual content, and encrypted messaging apps like Telegram and Signal for secure communication. Hashtags like #QuotaReformMovement and #StudentsAgainstQuota trended, forcing mainstream media to cover protests they might otherwise have ignored (Fig. 3.1).

The government's response followed predictable patterns: deploying police and ruling party youth wings to violently suppress protests, arresting student leaders, and eventually implementing a nationwide internet shutdown to prevent coordination and documentation. However, this strategy backfired spectacularly. Rather than quelling protests, the internet shutdown enraged broader populations, particularly middle-class urban families whose livelihoods depended on digital connectivity. The

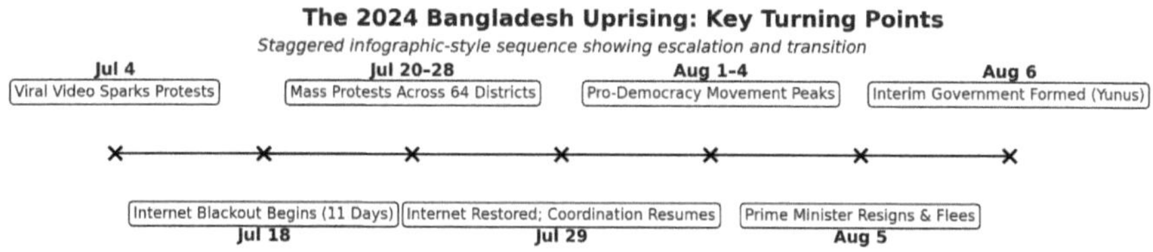

Fig. 3.1 Chronology of major events in the 2024 Bangladesh uprising—from the quota-reform trigger to the August 5 resignation. *Sources*: Ahmed (2024); Access Now (2024); Author analysis

shutdown also prevented the government from controlling narratives, as international media filled the information vacuum with reports of state violence (Ahmed, 2024).

Students adapted through remarkable resilience. When mobile internet was shut down, they used landline connections and WiFi. When those were restricted, they coordinated through traditional methods word of mouth, loudspeakers at mosques, and hand-written flyers. Diaspora Bangladeshis, particularly students abroad, used social media to amplify domestic voices internationally, applying diplomatic pressure on the Hasina government. Institutional Collapse and Social Media's Role: As protests intensified through July, institutional pillars of the regime began crumbling. Police forces, demoralized and stretched thin, became less willing to violently suppress protesters. Mid-level bureaucrats and even some ruling party members privately expressed sympathy for protesters. When the military declined to intervene decisively on Hasina's behalf, her position became untenable. Throughout this process, social media served crucial functions: coordinating protests across dozens of cities simultaneously documenting state violence (later used in international human rights forums), maintaining morale through shared experiences, and countering government propaganda. Even during internet blackouts, videos and images from Bangladesh continued appearing online, uploaded by those with access or carried across borders by travelers. When Hasina finally fled the country in early August 2024, social media exploded with celebrations, but also with immediate debates about transitional governance. Student leaders used social media to propose an interim government, coordinate with opposition parties, and maintain pressure for genuine democratic reforms rather than simple regime change. Institutional Responses and Reconstruction: Post-uprising, Bangladesh faces the challenge of rebuilding democratic institutions undermined during Hasina's rule. The interim government, including prominent civil society figures and student representatives, has used social media to communicate directly with citizens, seeking to build legitimacy and transparency. However, questions remain about regulating social media going forward how to prevent future authoritarian abuse of laws like the DSA while also combating misinformation and hate speech that could undermine fragile democratic progress. The uprising also forced institutions to confront their complicity in the former regime. The Election Commission, judiciary, and civil service all face demands for reform, with social media serving as a platform for debates about accountability and institutional reconstruction. Universities, which incubated the protest movement, are discussing curricular reforms to strengthen civic education and digital literacy. Significance for South Asian Context: Bangladesh's July 2024 uprising has profound implications for the broader South Asian region. It demonstrated that even highly repressive regimes with extensive surveillance and control capabilities can be overthrown through sustained, digitally coordinated popular mobilization. This has inspired activists in other countries particularly Pakistan and Sri Lanka while concerning governments throughout the region about their own vulnerabilities to youth-led digital movements (Fig. 3.2).

The uprising also highlighted generational divides in political engagement. Gen Z protesters roughly those born after 1997 showed fundamentally different political behaviors than previous generations. They organized horizontally rather than hierarchically, trusted peer networks over traditional leaders, communicated primarily through visual rather than text-based content, and proved remarkably resilient to state repression. Understanding this generational shift is essential for analyzing future political developments in South Asia.

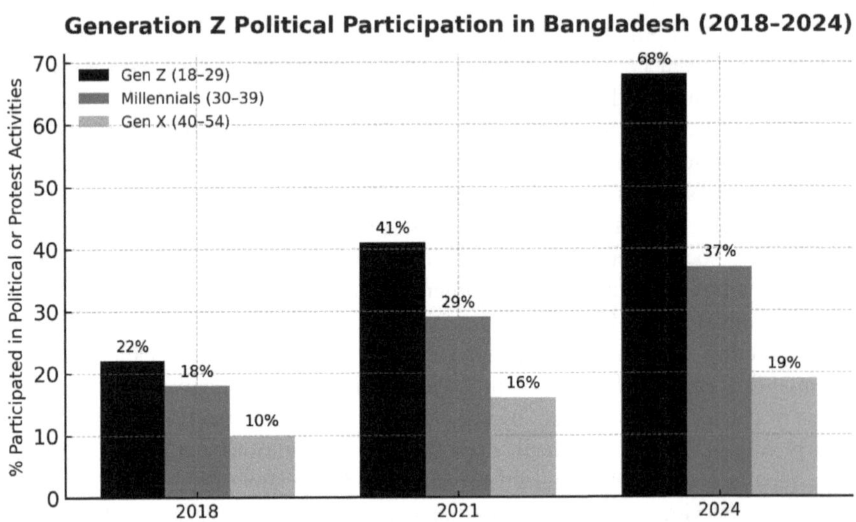

Year	Gen Z (18-29)	Millennials (30-39)	Gen X (40-54)
2018	22	18	10
2021	41	29	16
2024	68	37	19

Fig. 3.2 Comparative levels of online and street-level participation among generation Z and earlier cohorts, 2018–2024. *Sources* Ahmed and Haque (2024); IFES Youth Dataset

3.3.5 *Nepal: Gen Z and Political Transformation*

Nepal presents another compelling case of youth-led, digitally enabled political transformation, though with different dynamics than Bangladesh. Nepal's political instability following its 2015 constitution and the disillusionment with traditional parties created openings for new political actors, many emerging from social media organizing. Historical Context: Transition and Disappointment: Nepal's transition from monarchy to republic (2006–2008) raised hopes for democratic consolidation, but subsequent years brought political instability, corruption scandals, and glacial progress on constitutional promises. Traditional parties the Nepali Congress, Communist Party of Nepal (Unified Marxist Leninist), and various Maoist factions seemed to young Nepalis more interested in power-sharing than addressing unemployment, education quality, or economic development (Singh, 2020). This disillusionment particularly affected Nepal's large youth population (median age around 25), who increasingly engaged politically through social media rather than traditional party structures. Facebook and TikTok became spaces where young Nepalis critiqued political leaders, shared frustrations about economic opportunities, and experimented with new forms of political organization. Social Media and Youth Political Engagement: Unlike Bangladesh where social media organizing built toward a single explosive moment, Nepal's digital political engagement has been more diffuse, contributing to ongoing instability and pressure for systemic change. Youth-led campaigns on social media have targeted corruption (particularly in earthquake reconstruction funds after the 2015 earthquake), environmental degradation (opposing poorly planned infrastructure projects), and governance failures.

The Enough is Enough campaign in 2020–2021, primarily organized through social media, exemplified this trend. Young activists used Instagram and Twitter to document government failures, particularly in COVID-19 response, and called for leadership change. While not resulting in government overthrow like in Bangladesh, these campaigns contributed to the repeated collapse and reformation of governments in recent years. Rabi Lamichhane and the Rise of Social Media Politicians: The most striking development in Nepali politics has been the emergence of politicians who built followings primarily through social media. Rabi Lamichhane, a television journalist who gained fame through social media presence, founded the Rastriya Swatantra Party (RSP) in 2022, explicitly positioning it as an alternative to traditional parties and appealing directly to youth through digital platforms. Lamichhane's rapid rise becoming Home Minister within months of entering politics demonstrated social media's potential to disrupt traditional political hierarchies. His campaign relied heavily on YouTube videos, Facebook livestreams, and TikTok content, bypassing traditional media gatekeepers and party structures. While Lamichhane later faced corruption allegations himself (a reminder that social media-enabled politicians are not necessarily more ethical), his emergence reflects broader patterns of digital disruption in Nepali politics. Differential Impact Across Nepal: Social media's political influence varies significantly within Nepal based on geography, language, and ethnicity. Kathmandu Valley, with higher internet penetration and education levels, shows much stronger social media influence on politics than rural hill and mountain regions where connectivity remains limited. Additionally, most political social media content is in Nepali, excluding Madhesi, Tamang, and other linguistic minorities from full participation in digital political discourse.

3.3.6 Sri Lanka: Ethno-Religious Narratives and Facebook Shutdowns

Sri Lanka's experience with social media and democracy is heavily shaped by decades of ethnic civil war (1983–2009) between the Sinhalese-dominated government and Tamil separatists, as well as more recent tensions between the Buddhist majority and Muslim minority. Social media has become a vehicle for both ethnic nationalist mobilization and occasional violence. Anti-Muslim Violence and Facebook's Role: In 2018, Sri Lanka experienced anti-Muslim riots in the Kandy district, triggered by false rumors spreading on Facebook about Muslim business owners forcibly sterilizing Sinhalese workers and other conspiracies. Buddhist nationalist groups used Facebook to organize protests that escalated into violence, leading the government to temporarily shut down Facebook and other social media platforms a rare move for a democracy (Zubair, 2018). The shutdown sparked intense debate about balancing free expression against preventing violence. While the immediate violence decreased, critics argued the shutdown set dangerous precedents, threatened press freedom, and did not address underlying ethnic tensions or the structural factors making communities susceptible to misinformation. The shutdown also economically harmed small businesses dependent on social media for commerce and communication. Following the 2019 Easter Sunday bombings terrorist attacks on churches and hotels that killed over 250 people the government again shut down social media, claiming this would prevent further violence and the spread of false information. However, the shutdowns also prevented citizens from contacting loved ones, accessing news, or coordinating assistance, demonstrating the blunt nature of such measures. Platform Responses and Content Moderation Challenges: Sri Lanka pressured Facebook to improve content moderation in Sinhala and Tamil languages, arguing that the platform's inadequate moderation enabled hate speech and violence. Facebook acknowledged failures and committed to hiring more Sinhala and Tamil-speaking moderators, but challenges persist. The platforms' economic model prioritizes engagement, which can conflict with safety, and their global policy frameworks struggle to account for local contexts of ethnic conflict. Current Political Crisis and Social Media: Sri Lanka's recent economic collapse (2021–2022) and resulting political crisis saw social media playing crucial roles. Citizens organized protests through Facebook and Twitter, documenting government failures and demanding leadership change. The #GoHomeGota movement, targeting President Gotabaya Rajapaksa, originated on social media before manifesting in massive street protests. Rajapaksa's eventual resignation demonstrated social media's mobilization power even in contexts of severe economic distress.

3.3.7 *Understanding Differential Impact: Why Social Media Influence Varies Across South Asia*

Having examined specific country cases, we can now synthesize factors explaining why social media's influence on democracy varies across South Asia despite shared regional characteristics: State Capacity and Repressive Capabilities: Countries with stronger state capacity particularly sophisticated surveillance and security apparatus can more effectively control social media, though not without costs. India's relatively developed bureaucracy and Pakistan's military intelligence have greater technical capabilities for monitoring and manipulating social media than Nepal or Bangladesh (pre-2024). However, Bangladesh's July 2024 uprising demonstrated that even high state capacity cannot guarantee control when popular mobilization reaches critical thresholds. Civil Society Strength and Media Ecosystem: Countries with stronger independent civil society and more competitive media ecosystems show greater resilience against social media manipulation. India's vibrant civil society and relatively free press (despite recent deterioration) create more robust fact-checking and alternative information sources than Bangladesh under Hasina or Pakistan under military influence. Sri Lanka occupies a middle position, with civil society weakened by decades of conflict but gradually recovering. Internet Penetration and Digital Literacy: The proportion of the population with internet access and the level of digital literacy fundamentally shape social media's political impact. Higher connectivity means social media reaches more voters directly, but lower digital literacy increases vulnerability to misinformation. India's massive internet penetration means social media directly influences hundreds of millions of voters, while in less connected countries, social media influence filters through intermediaries, changing its character. Political Competition and Electoral Democracy: Social media's influence varies based on whether it operates within competitive electoral democracy, electoral autocracy, or military-influenced systems. In India's competitive multi-party system, social media serves as a battleground for competing narratives. In Bangladesh (until 2024), the Awami League's electoral dominance and control over institutions meant social media was primarily a space for opposition organizing and government propaganda, not genuine electoral competition. Pakistan's civil-military hybrid system creates unique dynamics where social media campaigns can be state-managed or military-influenced. Linguistic and Ethnic Diversity: The number of languages and ethnic groups fundamentally affects social media's influence. India's extraordinary linguistic diversity (hundreds of languages, dozens with significant speaker populations) creates fragmented digital public spheres where misinformation in one language may not translate to others, but also where content moderation is nearly impossible at scale. More linguistically homogeneous countries like Bangladesh (where Bengali dominates) have more unified digital public spheres, enabling both more effective national movements and more complete state control. Historical Context of Political Violence: Countries with more recent experiences of severe political violence Sri Lanka's civil war, Pakistan's terrorism challenges have populations more receptive to security justifications for social media restrictions. Meanwhile, Nepal's relatively peaceful recent history (despite the Maoist insurgency) means populations may be more resistant to censorship justified on security grounds. Youth Demographics and Economic Opportunities: While all South Asian countries have young populations, the specific economic opportunities (or lack thereof) available to youth shape how they engage politically through social media. Nepal's limited economic opportunities despite high education levels create frustration that fuels digital political engagement. Bangladesh's garment industry provides employment but at low wages, creating different economic grievances. These economic contexts shape what young people demand through social media organizing. International Connectivity and Diaspora Influence: Countries with large, politically engaged diasporas (Bangladesh, Pakistan, Sri Lanka) experience different social media dynamics than those with smaller or less politically active overseas populations. Diaspora communities can amplify domestic digital campaigns internationally, apply diplomatic pressure, and provide material support (including technical infrastructure like VPNs during shutdowns). This creates vulnerabilities for authoritarian governments like Hasina's, which could restrict domestic social media but not control diaspora communications. These factors interact in complex ways, explaining why social media facilitated youth-led uprisings in Bangladesh and contributed to instability in Nepal, while in India it has enhanced BJP dominance and in Pakistan it operates under military influence. Understanding these differential impacts is essential for developing effective policy responses, as solutions must be tailored to specific national contexts rather than assuming uniform patterns across South Asia. 3.6 Comparing the U.S. and Bangladesh: Contrasts in Scale, Language Barriers, Platform Usage, and Political Context.

3.3.8 The United States: High-Tech Disinformation and Polarized

Media Scale and Technological Sophistication: The U.S. social media environment is vast and technologically advanced, with massive user bases on platforms like Facebook, Twitter (X), and YouTube. Political operatives employ data analytics, microtargeting, and psychographic profiling, as seen in the Cambridge Analytica scandal (Cadwalladr & Graham-Harrison, 2018). Misinformation can be highly strategic, harnessing elaborate narratives or conspiracies like Pizzagate, QAnon, or election fraud claims that spread among ideologically aligned communities. Polarization and Echo Chambers: The U.S. media landscape is sharply polarized, with conservative and liberal outlets often reinforcing distinct worldviews (Sunstein, 2017). Social media algorithms amplify these cleavages, creating filter bubbles where users rarely encounter credible counter-arguments. During high-profile elections (e.g., 2016, 2020), rumors about vote rigging or candidate health circulated rapidly, sometimes boosted by influential political figures themselves. Regulatory Responses: The U.S. grapples with how to balance First Amendment protections against calls for greater oversight of social media. Section 230 of the Communications Decency Act protects platforms from liability over user-generated content, limiting the state's direct role in policing misinformation (Kosseff, 2019). Nonetheless, Congressional hearings and public pressure have forced major platforms to introduce fact-checking labels, ban certain false claims about COVID-19 or election integrity, and suspend high-profile individuals deemed to incite violence.

3.3.9 Bangladesh: Misinformation Intersecting with Low Digital Literacy and Government Control

Rapid Digital Expansion, Lower Literacy: Bangladesh's internet user base soared over the past decade, fueled by affordable smartphones and the government's "Digital Bangladesh" initiative. However, digital literacy remains uneven. Millions of new users lack the critical thinking skills to evaluate online content, making them vulnerable to sensational or misleading posts (Rahman & Tania, 2020b). Political parties exploit these gaps by circulating rumors that can quickly mobilize or intimidate communities. Role of WhatsApp and Facebook: While Facebook is the dominant platform for public political discussions, WhatsApp (and to a lesser extent Telegram) are crucial for private communication. Encrypted messaging fosters rumor dissemination, as it is difficult to trace or debunk false claims within closed groups. Politicians in Bangladesh are known to rely on cyber teams or "troll armies" to mass-report opposition content, spread rumors about rival candidates, and champion pro-government narratives (Ahmed, 2020). Regulatory Environment and the July 2024 Turning Point: The Bangladeshi government long wielded laws like the Digital Security Act (DSA) to arrest individuals for "defamatory" or "anti-state" social media posts (Human Rights Watch, 2021). Although officially justified as a measure against misinformation and cybercrime, critics argued that the DSA disproportionately targeted journalists and opposition activists, creating a climate of self-censorship rather than fostering a transparent information ecosystem. However, the July 2024 uprising demonstrated the limits of such repressive approaches. When the Hasina government attempted to suppress student-led protests through internet shutdowns and intensified use of the DSA, these tactics backfired spectacularly. Rather than quelling dissent, the digital blockade enraged broader populations and ultimately contributed to the regime's collapse. The post-uprising interim government now faces the challenge of reforming digital governance moving away from the DSA's repressive framework while still addressing legitimate concerns about misinformation and online hate speech (Ahmed, 2024). 3.6.3 Language and Cultural Dimensions Multilingual Contexts: In the U.S., English dominates national discourse, though Spanish and other languages have significant user bases. In Bangladesh, Bengali is the primary language, but smaller communities speak dialects that rarely see robust fact-checking resources. This linguistic complexity complicates the detection and debunking of rumors. Communal versus Partisan Cleavages: While U.S. misinformation frequently hinges on partisan divides, misinformation in Bangladesh often stirs communal sentiment targeting religious minorities or exploiting sectarian tensions for political gain (Rahman & Karim, 2020a). Both contexts reveal how social media manipulations feed off existing societal fault lines, but they manifest differently based on each country's cultural and historical backdrop. State Actor Involvement: In the U.S., foreign interference particularly from Russia played a significant role in 2016. In Bangladesh under Hasina, domestic government agencies and pro-government elements were more directly involved in controlling narratives, especially around election periods. Opposition parties also attempted to push their propaganda but faced greater legal risks and infrastructural disadvantages. The July 2024 uprising, however, demonstrated that sustained digital mobilization by citizens particularly youth could overcome even well-resourced state propaganda apparatus when political grievances reached critical thresholds.

3.3.10 Impact on Voter Behavior and Political Discourse

The U.S.: Misinformation fosters cynicism and mistrust in electoral outcomes, as a portion of the electorate may genuinely believe results were rigged or manipulated. This erodes faith in institutions like the electoral college and the judiciary. Media polarization intensifies these beliefs, making bipartisan consensus on policy even rarer. Bangladesh: While political cynicism has historically risen due to misinformation, the state's heavier hand under Hasina meant that misinformation was often overshadowed by censorship and intimidation. Voters struggled to sift through rumors about candidates, especially in rural or semi-literate settings, where local "information gatekeepers" shaped opinions. Communal violence could erupt when misinformation targeted religious communities, creating real-world consequences far beyond ballot boxes. The July 2024 events, however, revealed how social media could enable transformative political change despite or perhaps because of government repression. Youth activists used digital platforms to organize protests, document state violence, and maintain communication networks despite internet shutdowns. This demonstrated that in contexts of extreme political polarization and state repression, social media's mobilization potential could exceed its manipulation potential, though at considerable cost in terms of lives lost and social disruption. The challenge for post-uprising Bangladesh is whether the country can develop healthier digital political communication that harnesses social media's democratizing potential while mitigating its divisive and destabilizing effects.

3.4 Implications: Influence on Public Trust, Social Cohesion, and Electoral Integrity

3.4.1 Erosion of Public Trust in Institutions

At the heart of sustained misinformation lies a corrosive impact on trust citizens become uncertain about what to believe and who to trust (Lazer et al., 2018). Conspiratorial thinking can escalate as people feel official sources are biased or part of an "establishment conspiracy." Once trust in mainstream journalism or electoral commissions falters, unscrupulous actors find it easier to fill the vacuum with self-serving or divisive content. Voter Disillusionment: When rumors about vote rigging or election day violence circulate unchecked, portions of the electorate may lose confidence in the legitimacy of the process, potentially lowering voter turnout or acceptance of results. Media Skepticism: Users may adopt a cynical stance toward all news real or fake thus depriving them of reliable information necessary for informed decision-making.

3.4.2 Undermining Social Cohesion and Amplifying Violence

Misinformation can go beyond distorting political choices; it can incite communal violence or vigilantism when targeted groups are demonized. Across South Asia, false rumors about blasphemy, child kidnappings, or inter-ethnic conspiracies have led to mob attacks and killings (Arun, 2019). In the U.S., conspiracy-driven mobs were instrumental in the January 6th Capitol insurrection, highlighting how disinformation can catalyze real-world threats to democratic institutions (Zolan, 2021). Radicalization: Extremist groups leverage conspiracies to radicalize young recruits, framing out-groups or the government as existential threats. Online echo chambers accelerate radicalization by insulating believers from contradictory evidence (Boduszyński, 2020). Socioeconomic Impact: Periods of unrest can hamper economic stability, disrupt local businesses, and deter foreign investment outcomes particularly severe in lower-income regions already grappling with development challenges.

3.4.3 Electoral Integrity and Democratic Governance

Free and fair elections hinge on voters' access to accurate, diverse information regarding candidates and policies. Misinformation compromises electoral integrity by: Misleading Voters: Fabricated claims about a candidate's criminal history or false endorsements might sway marginal voters in critical constituencies. Suppressing Turnout: Rumors of violence or voter intimidation reduce turnout among vulnerable groups. Creating Post-Election Crises: Even after ballots are counted, allegations of fraud amplified by social media can delegitimize winners and destabilize governance, as seen in the U.S. following the 2020 election. In authoritarian or semi-authoritarian contexts, state-aligned media manipulate the narrative to favor incumbents. Yet, the existence of social media also provides a platform for dissenting voices unless it is heavily policed or censored under broad "anti-misinformation" laws, which ironically can be weaponized to quell legitimate criticism (Human Rights Watch, 2021).

3.4.4 Countering Disinformation: Stakeholder Efforts and Initiatives

Given these far-reaching implications, multiple stakeholders fact-checking organizations, NGOs, tech platforms, and governments endeavor to counter misinformation:

Fact-Checking Initiatives: Groups like PolitiFact in the U.S., BOOM

And Alt News in India, or BD Fact Check in Bangladesh systematically verify viral stories, producing easy-to-share refutations. However, these organizations often lack resources to match the volume and velocity of misinformation (Graves & Cherubini, 2016). NGO Campaigns and Media Literacy: Civil society and philanthropic bodies fund media literacy programs targeting youth and marginalized communities. Workshops, school curricula, and online toolkits teach participants to assess the credibility of sources, identify fake images, and avoid sensational "clickbait." Platform Regulations: Technology companies have introduced labeling systems, official "COVID-19 hubs," or election integrity initiatives to limit the spread of harmful content. Yet, critics argue platforms move too slowly, apply rules inconsistently, or fail to address the root cause of engagement-driven algorithms (Vaidhyanathan, 2018). Government Policies: Some governments consider legislation penalizing platforms for hosting false or harmful content, but these laws can be abused. Balancing free speech with the need for content moderation remains a contentious dilemma, especially in democracies with robust constitutional protections (Kosseff, 2019). International Collaborations: Organizations like the EU DisinfoLab, the International Fact-Checking Network (IFCN), and UNESCO's "MIL (Media and Information Literacy) Programme" advocate cross-border cooperation in knowledge-sharing and best practices to tackle disinformation. Global corporations, philanthropic foundations, and academic researchers increasingly coordinate their approaches to shape stronger institutional responses (Wardle & Derakhshan, 2017).

3.4.5 Institutional Implications and Democratic Resilience

The analysis of institutional responses to misinformation (Sect. 3.4) and regional case studies (Sect. 3.5) reveals profound implications for democratic institutions and governance. Social media has not simply added new communication channels to existing institutional structures but has fundamentally altered how institutions function and how their authority is perceived. Institutional Authority Under Challenge: Traditional sources of institutional authority election commissions announcing results, courts issuing rulings, fact-checkers verifying claims increasingly compete with alternative narratives circulating on social media. In environments of high polarization and low institutional trust, official pronouncements no longer settle debates but often intensify them. Citizens curate their own information ecosystems, selecting sources that confirm prior beliefs and dismissing institutional authorities as biased or corrupt. This authority erosion creates vicious cycles: as institutions lose credibility, they become less effective at fulfilling democratic functions, which further reduces public trust. Bangladesh's Election Commission under Hasina became so discredited that no official announcement could convince opposition supporters of election integrity, contributing to the legitimacy crisis that enabled the July 2024 uprising. Similarly, Pakistan's judiciary faces persistent questions about independence, with social media campaigns amplifying doubts regardless of court rulings' actual merits. Adaptation Capacity and Resource Constraints: Institutions vary enormously in their capacity to adapt to digital environments. Well-resourced organizations in developed democracies struggle with social media's scale and speed; those in developing countries with limited budgets and technical expertise face even steeper challenges. This creates exploitable vulnerabilities, as actors with resources to invest in digital manipulation can overwhelm under-resourced institutional defenses. The capacity gap also affects which institutions adapt successfully. Political parties even in resource-constrained environments often invest heavily in social media capabilities because electoral survival depends on it. Meanwhile, regulatory bodies, judiciaries, and election commissions may lack both resources and internal expertise to develop comparable capacities. This imbalance advantages political actors over supposedly neutral institutional referees, distorting democratic competition. Transparency Paradoxes and Information Overload: Social media promises transparency enabling citizens to monitor institutions and hold them accountable but simultaneously generates opacity through sheer information volume, misinformation prevalence, and algorithmic curation. Citizens may feel more informed while actually being more confused, having access to more information but less ability to process it critically. This paradox particularly affects developing democracies where populations have limited experience evaluating information critically. In Bangladesh, social media enabled unprecedented documentation of electoral irregularities in 2018, but also flooded information spaces with conflicting claims, making it difficult for ordinary citizens to distinguish genuine evidence from propaganda. The result was not enhanced transparency but confusion that actors with clearer narratives (in this case, the ruling party) could exploit. Implications for Democratic Quality and Stability:

The institutional transformations documented above carry profound implications for democratic quality. When citizens cannot trust election results, court rulings, or verified information, democratic governance's foundations crumble. When political parties prioritize viral manipulation over policy deliberation, democratic accountability suffers. When civil society organizations must dedicate resources to defending against digital attacks rather than advancing missions, democratic quality deteriorates. However, the picture is not uniformly bleak. Some institutions demonstrate impressive resilience and adaptation. India's fact-checking ecosystem, while imperfect, represents genuine institutional innovation. Bangladesh's youth activists developed sophisticated digital organizing capacities under repressive conditions, demonstrating civil society's resilience. These examples suggest that while social media challenges institutions profoundly, institutional collapse is not inevitable adaptation is possible when resources, will, and favorable conditions align. The key question is whether institutional adaptation can occur quickly and comprehensively enough to preserve democratic integrity against rapidly evolving digital threats. The answer appears to depend heavily on context: countries with stronger rule of law, independent media, robust civil society, and competitive politics show more institutional resilience than those lacking these features. This suggests that strengthening traditional democratic institutions even as they adapt to digital challenges remains essential, rather than assuming that social media itself will solve governance problems.

3.4.6 *Institutional Implications and Democratic Resilience*

The analysis of institutional responses to misinformation (Sect. 3.4) and regional case studies (Sect. 3.5) reveals profound implications for democratic institutions and governance. Social media has not simply added new communication channels to existing institutional structures but has fundamentally altered how institutions function and how their authority is perceived. Institutional Authority Under Challenge: Traditional sources of institutional authority election commissions announcing results, courts issuing rulings, fact-checkers verifying claims increasingly compete with alternative narratives circulating on social media. In environments of high polarization and low institutional trust, official pronouncements no longer settle debates but often intensify them. Citizens curate their own information ecosystems, selecting sources that confirm prior beliefs and dismissing institutional authorities as biased or corrupt. This authority erosion creates vicious cycles: as institutions lose credibility, they become less effective at fulfilling democratic functions, which further reduces public trust. Bangladesh's Election Commission under Hasina became so discredited that no official announcement could convince opposition supporters of election integrity, contributing to the legitimacy crisis that enabled the July 2024uprising. Similarly, Pakistan's judiciary faces persistent questions about independence, with social media campaigns amplifying doubts regardless of court rulings' actual merits. Adaptation Capacity and Resource Constraints: Institutions vary enormously in their capacity to adapt to digital environments. Well-resourced organizations in developed democracies struggle with social media's scale and speed; those in developing countries with limited budgets and technical expertise face even steeper challenges. This creates exploitable vulnerabilities, as actors with resources to invest in digital manipulation can overwhelm under-resourced institutional defenses. The capacity gap also affects which institutions adapt successfully. Political parties even in resource-constrained environments often invest heavily in social media capabilities because electoral survival depends on it. Meanwhile, regulatory bodies, judiciaries, and election commissions may lack both resources and internal expertise to develop comparable capacities. This imbalance advantages political actors over supposedly neutral institutional referees, distorting democratic competition. Transparency Paradoxes and Information Overload: Social media promises transparency enabling citizens to monitor institutions and hold them accountable but simultaneously generates opacity through sheer information volume, misinformation prevalence, and algorithmic curation. Citizens may feel more informed while actually being more confused, having access to more information but less ability to process it critically. This paradox particularly affects developing democracies where populations have limited experience evaluating information critically. In Bangladesh, social media enabled unprecedented documentation of electoral irregularities in 2018, but also flooded information spaces with conflicting claims, making it difficult for ordinary citizens to distinguish genuine evidence from propaganda. The result was not enhanced transparency but confusion that actors with clearer narratives (in this case, the ruling party) could exploit. Implications for Democratic Quality and Stability: The institutional transformations documented above carry profound implications for democratic quality. When citizens cannot trust election results, court rulings, or verified information, democratic governance's foundations crumble. When political parties prioritize viral manipulation over policy deliberation, democratic accountability suffers. When civil society organizations must dedicate resources to defending against digital attacks rather than advancing missions, democratic quality deteriorates. However, the picture is not uniformly bleak. Some institutions demonstrate impressive resilience and adaptation. India's fact-checking ecosystem, while imperfect, represents genuine institutional innovation. Bangladesh's youth activists developed sophisticated digital organizing capacities under repressive conditions, demonstrating civil society's resilience. These examples suggest that while social media challenges institutions profoundly, institutional collapse is not inevitable adaptation is possible when resources, will, and favorable conditions align. The key question is whether institutional adaptation can occur quickly and comprehensively enough to preserve democratic integrity against rapidly evolving digital threats. The answer appears to depend heavily on context: countries with stronger rule of law, independent media, robust civil society, and competitive politics show more institutional resilience than those lacking these features. This suggests that strengthening traditional democratic institutions even as they adapt to digital challenges remains essential, rather than assuming that social media itself will solve governance problems.

3.5 Discussion

This chapter has examined misinformation and manipulation through conceptual, mechanical, institutional, and regional analytical frameworks, revealing how digital platforms fundamentally transform democratic politics beyond simply adding new communication channels. Three core insights emerge from this multi-level analysis. First, misinformation operates through interconnected systems rather than isolated incidents. The conceptual distinctions between misinformation, disinformation, and propaganda while analytically useful blur in practice as content circulates through complex digital ecosystems. What begins as unintentional misinformation can be weaponized into deliberate disinformation; propaganda campaigns exploit organic misinformation to amplify their reach. The mechanisms of dissemination algorithms optimizing for engagement, bots creating artificial popularity, troll armies coordinating attacks, and deepfakes undermining trust in evidence function synergistically to create environments where falsehoods systematically outcompete truth. Understanding these interconnections is essential; interventions targeting single mechanisms while ignoring systemic dynamics are unlikely to succeed. Second, institutional transformation represents misinformation's most profound democratic threat. While individual instances of false information matter, the systematic erosion of institutional authority poses deeper challenges. Election management bodies, courts, regulatory agencies, political parties, and civil society organizations all find their traditional functions disrupted by social media dynamics. Institutional pronouncements no longer settle debates but often intensify them. Authority derives increasingly from viral popularity rather than expertise or democratic legitimacy. Political parties have evolved from organizations competing within shared rules into sophisticated manipulators employing cyber armies to attack opponents and spread propaganda. These transformations create vicious cycles: weakened institutions become less capable of countering misinformation, further reducing public trust and institutional effectiveness. Third, context fundamentally shapes outcomes. The comparative analysis across South Asia and between the U.S. and Bangladesh demonstrates that identical technologies produce vastly different political dynamics depending on institutional strength, state capacity, civil society development, media ecosystem maturity, linguistic diversity, and historical legacies. India's WhatsApp lynchings, Pakistan's military-influenced digital propaganda, Bangladesh's youth-led uprising, and Nepal's social media politicians reflect both regional commonalities and country-specific configurations. This context-dependency challenges universalist assumptions about social media's effects and suggests that effective responses must be tailored to specific political, economic, and cultural environments. Theoretical Contributions and Limitations: This chapter advances understanding of digital misinformation by integrating multiple analytical levels often examined separately. The institutional focus extends beyond typical analyses of individual-level effects or platform mechanisms to examine how democratic institutions themselves transform. The comparative regional analysis reveals patterns obscured by single-country studies, while also demonstrating important variations that cross-national research risks oversimplifying. However, several limitations merit acknowledgment. The analysis relies primarily on secondary literature and documented cases rather than original empirical research on misinformation networks or platform algorithms. The institutional analysis, while comprehensive in scope, would benefit from deeper examination of specific institutional responses and their effectiveness. The South Asian focus, while providing regional depth, limits generalizability to other developing regions like Africa or Latin America with different historical trajectories and digital landscapes. Finally, the rapidly evolving nature of digital technologies means that mechanisms and impacts documented here may shift as platforms, regulations, and user behaviors change. Implications for Subsequent Chapters: The patterns identified here establish foundations for analyses in subsequent chapters. Chapter 4's examination of regulatory frameworks must grapple with the institutional vulnerabilities and context-dependencies documented here. The case studies in Chapts. 5–7 will illustrate how these theoretical dynamics manifest in specific electoral contexts. The comparative analysis in Chap. 8 will build on the differential impacts across contexts identified here. Most critically, Chap. 9's policy recommendations must address the systemic nature of misinformation, the institutional transformations it produces, and the necessity of context-appropriate interventions rather than one-size-fits-all solutions. The analysis reveals that misinformation's challenge is not primarily technological but fundamentally political and institutional. Social media platforms amplify and accelerate underlying dynamics partisan polarization, institutional weakness, social cleavages, information poverty but do not create them. This suggests that purely technological solutions better algorithms, improved fact-checking tools while valuable, cannot alone restore democratic integrity. Addressing misinformation requires strengthening democratic institutions, rebuilding public trust, bridging social divisions, and ensuring that technological development serves rather than undermines democratic values. How democracies navigate these challenges will determine whether digital platforms ultimately enhance or erode democratic governance.

3.6 Conclusion

Social media has fundamentally transformed democratic politics, creating unprecedented opportunities for participation while simultaneously threatening the institutional foundations that democracies require to function. This chapter has demonstrated that misinformation's challenge extends far beyond individual false claims to encompass the systematic erosion of institutional authority, the fragmentation of shared informational spaces, and the weaponization of digital platforms against democratic processes. The South Asian cases examined here particularly Bangladesh's July 2024 uprising reveal that social media can enable both democratic liberation and authoritarian control, with outcomes depending on complex interactions between technology, institutions, and human agency. When youth activists used digital platforms to coordinate protests despite internet shutdowns and state violence, they demonstrated social media's potential to empower citizens against repressive regimes. Yet these same platforms had previously enabled the Hasina government's propaganda and surveillance, illustrating how identical technologies can serve opposite purposes depending on who controls them and how they are used. This duality captures the essential challenge facing democracies worldwide: how to preserve the benefits of digital communication unprecedented access to information, new forms of political participation, rapid mobilization capabilities while mitigating the harms of misinformation, manipulation, and institutional erosion. There are no easy answers, no purely technological solutions, no universal policies that will work across all contexts. What is clear is that meeting this challenge requires coordinated effort across multiple levels of society. Individuals must develop critical media literacy. Institutions must build capacity to operate effectively in digital environments. Platforms must prioritize accuracy over engagement. Governments must craft regulations that prevent manipulation without enabling authoritarian abuse. Civil society must maintain its watchdog functions despite digital harassment. International cooperation must address transnational threats while respecting sovereignty. Most fundamentally, preserving democracy in the digital age requires recognizing that this is not a temporary crisis to be solved but an ongoing condition to be managed. Digital platforms will continue evolving; manipulation tactics will become more sophisticated; new challenges will emerge. Democratic resilience requires not just responding to current threats but developing adaptive capacity to address future ones. The coming years will determine whether democracies can develop effective responses that preserve both the benefits and the institutional foundations necessary for democratic governance. As this chapter has shown, the outcome is far from predetermined. Social media can enable democratic transformation or authoritarian consolidation, informed deliberation or misinformed polarization, institutional strengthening or institutional collapse. Which outcome prevails depends on choices made by individuals, institutions, and societies choices that require understanding the complex dynamics this chapter has sought to illuminate. The work of building democratic resilience for the digital age has only just begun.

Case Study

The 2016 U.S. presidential election and India's 2019 Lok Sabha elections both exemplify how misinformation operates at industrial scale. In the United States, Russian-linked troll farms deployed sophisticated bots to amplify divisive content, exploiting Facebook's engagement-driven algorithms. Meanwhile, in India, WhatsApp became the vector for deadly rumors—false claims about child kidnappers led to mob lynchings in rural communities with limited digital literacy.

Bangladesh faces similar challenges but with distinct characteristics: language barriers limit fact-checking resources, lower literacy rates increase vulnerability to manipulated content, and state actors often blur the line between countering misinformation and suppressing legitimate dissent. The 2018 elections saw coordinated troll armies attacking opposition figures, while deepfake technology began emerging as a tool for political smear campaigns.

Questions for Discussion

1. How do misinformation dissemination mechanisms differ between text-based platforms (Twitter/X) and encrypted messaging apps (WhatsApp)?
2. What role do linguistic and literacy barriers play in misinformation susceptibility in multilingual democracies?
3. Should social media platforms bear legal liability for user-generated misinformation, or does this threaten free speech?
4. How can developing nations build fact-checking capacity when journalism itself faces economic and political constraints?
5. What distinguishes state propaganda from decentralized misinformation, and why does this distinction matter for regulatory responses?

Food for Thought

If virality rewards emotional intensity over factual accuracy, and human psychology favors confirmation bias, can any intervention truly counteract misinformation's structural advantages?

What We Learn

Misinformation thrives in the gaps between technological capability and societal adaptation. Combating it requires multi-stakeholder approaches: platform transparency, regulatory frameworks that balance free expression with accountability, media literacy education, and robust independent journalism. The challenge is particularly acute in developing democracies where limited resources and authoritarian tendencies can transform anti-misinformation efforts into tools of censorship.

References

Ahmed, F. (2020). Troll armies in emerging democracies: Bangladesh's online partisans. *Journal of Political Communication, 15*(3), 210–228.

Arun, C. (2019). On WhatsApp, rumors, and lynchings: the missing regulatory framework in India. *Economic and Political Weekly, 54*(6), 47–55.

Banaji, S., & Bhat, R. (2020). *WhatsApp vigilantes: Fake news and right-wing propaganda in India.* Oxford University Press.

Boduszyński, M. P. (2020). Digital radicalization in the Balkans: The case of Bosnia and Herzegovina. *Journal of Southeast European and Black Sea Studies, 20(2), 267–285.*

Bradshaw, S., & Howard, P. N. (2019). The global disinformation order: 2019 global inventory of organised social media manipulation. Oxford Internet Institute Working Paper. https://comprop.oii.ox.ac.uk/research/cybertroops2019/

Cadwalladr, C., & Graham-Harrison, E. (2018, March 17). Revealed: 50 million Facebook profiles harvested for Cambridge Analytica in major data breach. *The Guardian.* https://www.theguardian.com/news/2018/mar/17/cambridge-analytica-facebook-influence-us-election

Chakravartty, P., & Roy, S. (2015). Mr. Modi goes to Delhi: Mediated populism and the 2014 Indian elections. *Television & New Media, 16*(4), 311–322.

Gillespie, T. (2018). *Custodians of the internet: Platforms, content moderation, and the hidden decisions that shape social media.* Yale University Press.

Graves, L., & Cherubini, F. (2016). The rise of fact-checking sites in Europe. Reuters Institute for the Study of Journalism, University of Oxford.

Human Rights Watch. (2021). "*Muzzling dissent": The return of Bangladesh's repressive Information and Communication Technology Act in a new guise.* https://www.hrw.org/report/2021/03/31/muzzling-dissent/return-bangladeshs-repressive-information-and-communication

Kosseff, J. (2019). *The twenty-six words that created the internet.* Cornell University Press.

Kreiss, D. (2016). *Prototype politics: Technology-intensive campaigning and the data of democracy.* Oxford University Press.

Lazer, D. M. J., Baum, M. A., Benkler, Y., Berinsky, A. J., Greenhill, K. M., Menczer, F., Metzger, M. J., Nyhan, B., Pennycook, G., Rothschild, D., Schudson, M., Sloman, S. A., Sunstein, C. R., Thorson, E. A., Watts, D. J., & Zittrain, J. L. (2018). The science of fake news. *Science, 359*(6380), 1094 1096. https://doi.org/10.1126/science.aao2998

Nyhan, B., & Reifler, J. (2010). When corrections fail: The persistence of political misperceptions. *Political Behavior, 32*(2), 303–330.

Persily, N. (2017). Can democracy survive the internet? *Journal of Democracy, 28*(2), 63–76.

Rahman, T., & Karim, L. (2020a). Intersecting religion and politics in digital spaces: Communal violence in Bangladesh. *Social Media & Conflict Journal, 5*(1), 41–62.

Rahman, T., & Tania, T. (2020b). Rural connectivity and digital divide: A comparative analysis of north and south Bangladesh. *Telecom Policy Studies, 12*(2), 56–70.

Shahzad, M. (2021). Digital dynamics of electoral competition in Pakistan: Narratives, propaganda, and voter behavior. *South Asia Digital Quarterly, 8*(1), 71–90.

Silverman, C. (2020, November 12). CISA calls 2020 election "the most secure in American history" as Trump continues to claim fraud. *BuzzFeed News.* https://www.buzzfeednews.com/article/craigsilverman/cisa-most-secure-election-trump-fraud-claims

Singh, R. (2020). Social media misinformation in Nepal: A study of ethnic and political narratives. *Nepal Media Review, 4*(2), 112–132.

Sunstein, C. R. (2017). *#Republic: Divided democracy in the age of social media.* Princeton University Press.

Vaidhyanathan, S. (2018). *Antisocial media: How Facebook disconnects us and undermines democracy.* Oxford University Press.

Voltmer, K. (2013). *The media in transitional democracies.* Polity Press.

Wardle, C., & Derakhshan, H. (2017). *Information disorder: Toward an interdisciplinary framework for research and policy making.* Council of Europe. https://rm.coe.int/information-disorder-toward-an-interdisciplinary-framework-for-researc/168076277c

Zia, A. (2018). Digital activism in Pakistan: Censorship and resistance. *Journal of Political Communication, 14*(2), 112–132.

Zolan, K. (2021, January 8). Social media coordination in the Capitol riot. *Politico.* https://www.politico.com/news/2021/01/08/social-media-capitol-riot-456940

Zubair, A. (2018). Sri Lanka's Facebook ban: A digital experiment in hate speech regulation. *South Asian Human Rights Review, 2*(3), 89–102.

Chapter 4
Regulatory Frameworks Censorship, Algorithmic Governance, and Civil Liberties

4.1 Introduction

Over the past two decades, the exponential rise of digital communication has transformed the way people interact, disseminate information, and participate in political and economic life. Social media platforms have evolved from niche networking sites into vast, influential ecosystems that underpin the global flow of information. In this context, governments and corporate actors alike are forced to address a host of challenges related to misinformation, hate speech, extremist content, and the potential for digital propaganda. The complex interplay between state regulation, corporate self-governance, and individual rights has given rise to an ongoing debate: How can societies protect national security and public order while safeguarding civil liberties such as freedom of expression, privacy, and political dissent?

This chapter examines the regulatory frameworks that have emerged to manage these tensions, exploring their effectiveness and shortcomings in both established and emerging democracies. By comparing governmental approaches in countries such as the United States and Bangladesh, we reveal contrasting regulatory philosophies. In the United States, a long tradition of free speech protection has fostered a model that emphasizes corporate self-regulation and minimal direct state interference with Section 230 of the Communications Decency Act often cited as a cornerstone of this approach. In contrast, emerging democracies like Bangladesh have implemented more interventionist policies, such as the Digital Security Act (DSA), which many critics argue provides broad discretionary power to state authorities and poses significant risks to civil liberties.

In addition to contrasting governmental approaches, this chapter delves into the challenges posed by algorithmic governance. Digital platforms now rely on complex, opaque algorithms to curate content, moderate discussions, and target political advertisements. These algorithms can inadvertently or sometimes deliberately amplify bias, misinform voters, and shape public discourse in subtle yet profound ways. Calls for increased transparency, independent audits, and accountability mechanisms

B. Ahmed and R. J. Samir, *Beyond Ballot Boxes*,
https://doi.org/10.1007/978-3-032-27350-5_4

have grown louder as civil society groups demand to know how these "black boxes" operate and influence the digital public sphere.

Another critical area of concern is the balance between national security imperatives and the protection of civil liberties. In many states, governments argue that robust digital regulation, including surveillance measures and content restrictions, is necessary to combat terrorism, cybercrime, and extremist propaganda. However, such measures often encroach on fundamental freedoms, raising serious questions about the proper limits of state power and the preservation of open discourse. Moreover, as international controversies have demonstrated from the Snowden revelations in the United States to widespread protests in Bangladesh over internet shutdowns such regulatory tools may be misused to stifle dissent and control the narrative rather than to genuinely safeguard public order.

Finally, given the transnational nature of digital communications, international organizations and global watchdogs have emerged as influential players in shaping digital regulatory norms. Institutions like UNESCO, Reporters Without Borders, and various regional initiatives strive to establish standards that balance state security, corporate responsibility, and individual rights. Their guidelines and reports provide valuable benchmarks, yet they often clash with the political and economic realities of individual nations.

This chapter is structured into four major sections. First, we compare governmental approaches in the United States and Bangladesh, examining how differing political traditions and legal frameworks produce divergent outcomes. Next, we explore the realm of algorithmic governance, discussing the growing demands for transparency and independent oversight. We then address civil liberties concerns, analyzing the inherent tensions between state security measures and free expression. Finally, we review the role of international organizations and global watchdogs in shaping and enforcing digital rights. Throughout, we include real-world examples and in-depth analysis to illustrate how these regulatory frameworks function and sometimes malfunction in practice.

4.2 Governmental Approaches in Contrasting Contexts

4.2.1 The United States: Self-regulation, Section 230, and the Free Speech Paradigm

The United States has long been considered a bastion of free expression, a principle enshrined in the First Amendment. This legacy has heavily influenced its approach to digital regulation. Rather than imposing stringent state controls on online content, U.S. policymakers have largely opted for a model of self-regulation, wherein digital platforms are given broad discretion over the content they host. A critical legal instrument supporting this model is Section 230 of the Communications Decency

Act (1996), which grants online services immunity from liability for user-generated content (Kosseff, 2019).

Section 230 has been credited with fostering the explosive growth of user-generated content and the rise of social media giants like Facebook, Twitter (now X), and YouTube (Benkler, Faris, & Roberts, 2018). The rationale behind this protection is that it enables platforms to moderate content without fear of constant litigation, thus encouraging innovation and free expression. However, critics argue that Section 230 now shields these companies from accountability, even when their algorithms amplify harmful or polarizing content. For example, during the 2016 U.S. presidential election, critics pointed to the spread of extremist propaganda and disinformation as evidence that unchecked algorithms could distort democratic discourse (Benkler et al., 2018). Subsequent controversies such as the dissemination of conspiracy theories during the COVID-19 pandemic and the coordination of violent protests have further fueled calls for reform.

Despite growing bipartisan pressure for amending Section 230, attempts to modify or repeal its broad protections have met with fierce resistance from the tech industry and civil libertarians. Lawmakers from both sides of the aisle remain divided: some view the law as an essential enabler of online speech and innovation, while others see it as a loophole that allows harmful content to flourish unchecked. Consequently, U.S. policy in this area remains largely decentralized, with regulatory initiatives at the state level and ad hoc federal actions serving as the main corrective mechanisms. The debate over Section 230 encapsulates the broader tension in U.S. digital policy: how to balance the benefits of free expression and innovation against the need for accountability and the mitigation of online harm.

Beyond regulatory frameworks, social media has fundamentally altered how U.S. democratic institutions operate. Congressional oversight committees, initially reactive to tech industry developments, have been forced to build technical expertise to understand algorithmic systems. The 2018 Facebook hearings revealed significant institutional knowledge gaps, with some legislators asking questions that demonstrated fundamental misunderstanding of digital platforms (Kreiss & McGregor, 2019). This institutional learning curve has been steep but necessary, as Congress now regularly engages with platform architecture, content moderation systems, and algorithmic accountability in ways unimaginable a decade ago. The Federal Election Commission similarly struggles to adapt campaign finance regulations to digital advertising, where micro-targeted ads can circumvent traditional disclosure requirements, revealing how institutional structures designed for traditional media cannot adequately govern the speed, scale, and opacity of digital political communication.

4.2.2 *Institutional Transformation: How Social Media Reshapes Democratic Structures*

4.2.2.1 Electoral Management Bodies and the Digital Challenge

Electoral institutions worldwide face unprecedented challenges as social media platforms become primary arenas for political mobilization. In the United States, the Federal Election Commission (FEC) has struggled to adapt campaign finance regulations to digital advertising, where micro-targeted ads can circumvent traditional disclosure requirements (Kreiss & McGregor, 2019). The FEC's limited resources and technological capacity have left significant gaps in monitoring online political spending, particularly on platforms like Facebook and Twitter that operate globally but are accessed locally.

In Bangladesh, the Election Commission (EC) has attempted to establish digital monitoring units to track online campaign activities during election periods. However, these units face severe resource constraints and lack the technical sophistication to effectively monitor encrypted messaging platforms like WhatsApp and Telegram, where much political coordination occurs. The EC's reliance on manual reporting by citizens and journalists means that most violations go undetected until after electoral damage has occurred. Moreover, the EC's limited independence from the ruling government raises questions about selective enforcement, particularly when social media content favors incumbent parties.

Comparative Analysis: While U.S. electoral institutions operate with greater independence and resources, both contexts reveal a common pattern: institutional structures designed for traditional media cannot adequately govern the speed, scale, and opacity of digital political communication. This institutional lag creates opportunities for manipulation and undermines electoral integrity.

4.2.2.2 Judicial Systems Confronting Digital Evidence and Cyber Cases

Courts in both developed and developing democracies are grappling with digital content as evidence, jurisdictional questions about online speech, and the interpretation of laws written before social media existed. In the United States, landmark cases involving Section 230 immunity have forced courts to balance platform liability with free speech protections, often with inconsistent results across different circuit courts (Goldman, 2020). The Supreme Court has been reluctant to significantly narrow Section 230, recognizing the potential chilling effect on online expression, yet lower courts struggle with applying decades-old precedent to AI-driven content moderation.

Bangladesh's judiciary faces different but equally significant challenges. Under the Digital Security Act, courts process hundreds of cases involving social media posts, yet judges often lack training in digital forensics or understanding of online context (Ahmed, 2020). The admissibility of screenshots as evidence, the authentication of WhatsApp messages, and the interpretation of satirical content have led to controversial convictions that critics argue reflect judicial unfamiliarity with digital communication norms. Furthermore, the DSA's provisions allowing arrest without warrant have overwhelmed the judicial system, creating backlogs that delay justice for both complainants and defendants.

Social Media's Institutional Impact: The volume and velocity of digital cases have transformed judicial workloads, forcing courts to develop new protocols for evidence handling. However, neither U.S. nor Bangladeshi courts have fully institutionalized standards for digital evidence, leading to inconsistent rulings that undermine legal certainty.

4.2.2.3 Legislative Bodies and Social Media-Driven Policy Pressures

Parliaments and congresses worldwide face intense social media pressure that shapes legislative agendas and constrains policy deliberation. In the U.S. Congress, viral social media campaigns can rapidly mobilize public opinion, forcing legislators to respond to trending topics rather than engaging in thoughtful policy development (Kreiss, 2016). Congressional hearings on big tech companies have become highly publicized events, with legislators performing for social media audiences rather than conducting rigorous oversight. Twitter threads and Facebook posts by individual members of Congress increasingly substitute for formal legislative communication, fragmenting institutional voice and accountability.

In Bangladesh, the Jatiya Sangsad (National Parliament) experiences social media influence differently but with similar institutional consequences. Government MPs use Facebook extensively to promote development projects and counter opposition narratives, while opposition MPs when permitted to speak leverage social media to amplify their parliamentary speeches to audiences far beyond those watching parliamentary proceedings. However, the DSA has created a chilling effect on parliamentary debate itself, as MPs risk prosecution for statements made in parliament if those statements are then shared on social media and deemed defamatory or anti-state.

Institutional Evolution: Social media has democratized legislative communication but simultaneously weakened institutional coherence. The shift from collective party positions to individual member branding undermines party discipline and makes coalition-building more difficult. Both U.S. and Bangladeshi legislatures struggle to maintain institutional authority when individual politicians can circumvent party structures through direct social media engagement with constituents.

4.2.2.4 Traditional Media Institutions: Adaptation and Decline

Legacy media institutions newspapers, television networks, radio stations have been profoundly disrupted by social media platforms that now dominate information distribution. In the United States, newspaper readership has plummeted as Facebook and

Twitter became primary news sources for most Americans (Pew Research Center, 2021). This shift has decimated advertising revenue, forcing newsroom closures and reducing investigative journalism capacity. Television news networks increasingly chase social media trends, with producers monitoring Twitter to identify "breaking" stories, often at the expense of substantive reporting.

Bangladesh's traditional media landscape has undergone similar upheaval. Print newspapers that once dominated political discourse now struggle to compete with instant social media updates. Television news channels, while still influential, increasingly source their content from viral social media posts rather than conducting original reporting. This has created a dangerous feedback loop: traditional media legitimizes unverified social media content by broadcasting it, which then lends the content credibility that spreads further on social media.

Institutional Consequences: The economic decline of traditional media has weakened institutional checks on political power. As newsrooms shrink, investigative capacity diminishes, reducing accountability journalism precisely when social media makes misinformation easier to spread. In Bangladesh, where press freedom is already constrained, this creates an information vacuum filled by partisan social media content.

4.2.2.5 Civil Society Organizations in the Digital Public Sphere

Non-governmental organizations, advocacy groups, and civil society watchdogs have both benefited from and been challenged by social media's rise. In the United States, organizations like the ACLU and Electronic Frontier Foundation use social media for advocacy campaigns, reaching millions with petitions and mobilization efforts that would have required enormous traditional media budgets (EFF, 2021). However, these same organizations struggle to combat coordinated disinformation campaigns and astroturfing that mimic genuine grassroots activism.

Bangladeshi civil society operates under greater constraints but has similarly adopted social media for advocacy. Human rights organizations document DSA abuses through Facebook posts and Twitter threads, often reaching international audiences when domestic media coverage is restricted (Human Rights Watch, 2021). Youth-led movements, such as the 2018 road safety protests, demonstrated social media's power to coordinate civil society action independent of traditional organizational structures. Yet the government's subsequent crackdown on protest organizers, facilitated by social media surveillance, revealed the double-edged nature of digital activism.

Institutional Adaptation: Civil society organizations must now master digital advocacy while protecting themselves from digital surveillance and harassment. This has led to institutional changes: hiring social media managers, developing digital security protocols, and creating encrypted communication channels. The institutional capacity gap between well-resourced international NGOs and local grassroots organizations has widened in the digital age.

4.2.2.6 Political Party Structures and Digital Transformation

Political parties foundational institutions of democratic systems have undergone structural changes driven by social media. In the United States, both major parties have created sophisticated digital operations, with dedicated social media teams, data analytics divisions, and influencer outreach programs (Kreiss, 2016). However, this has also decentralized party power, as individual candidates can build personal brands and fundraising networks independent of party infrastructure. The rise of outsider candidates like Donald Trump in 2016, whose Twitter presence substituted for traditional party organizing, exemplified this institutional disruption.

Bangladesh's political parties show similar patterns adapted to local context. The Awami League has established "cyber teams" that operate as semi-formal party structures, monitoring social media, coordinating pro-government messaging, and reporting opposition content for DSA prosecution (Ahmed, 2020). The BNP, facing resource constraints and government pressure, has relied more on diaspora-based social media networks operating outside Bangladesh's jurisdiction. Jamaat-e-Islami has leveraged encrypted messaging apps to coordinate religious mobilization while avoiding direct government surveillance (KII #2, 2025).

Institutional Impact: Social media has simultaneously strengthened and weakened party institutions. Parties can mobilize supporters more efficiently but struggle to maintain message discipline when members can communicate directly with constituents. The shift from hierarchical party structures to networked digital operations represents a fundamental transformation in how political parties function as democratic institutions.

4.3 Algorithmic Governance: Transparency, Accountability, and the Search for Solutions

4.3.1 The Rise of Algorithmic Governance and Its Implications

Digital platforms have become much more than passive repositories of user-generated content; they now actively curate and shape the information landscape through sophisticated algorithms. These algorithms determine which posts appear in a user's feed, which videos are recommended, and how advertisements are targeted. As these processes are largely automated and proprietary, they operate as "black boxes" that are difficult for external observers to scrutinize (Vaidhyanathan, 2018). Although companies assert that their algorithms are designed to maximize user engagement and provide personalized experiences, growing evidence suggests that these systems often favor sensational, polarizing, or otherwise divisive content.

Algorithmic bias has emerged as a critical issue in digital governance. Studies indicate that recommendation systems on platforms like YouTube can inadvertently radicalize users by suggesting increasingly extreme content (O'Donovan et al., 2019). Similarly, Facebook's newsfeed algorithm has been criticized for promoting posts that generate strong emotional reactions, which often means content that is misleading, inflammatory, or divisive. This tendency exacerbates echo chambers and filter bubbles, wherein users are exposed predominantly to information that confirms their preexisting beliefs (Pariser, 2011).

The challenge is compounded by the lack of transparency. Platform companies guard their algorithms as trade secrets, making it nearly impossible for researchers, journalists, or regulators to assess how content is prioritized or why certain narratives gain prominence. Although some platforms have introduced transparency reports and external audits, these initiatives are often criticized as insufficient. Without genuine accountability, the potential for algorithmic manipulation whether intentional or accidental remains a significant threat to democratic processes.

4.3.2 *Demands for Transparency and Independent Audits*

Civil society groups, academics, and some policymakers have increasingly called for mandatory transparency in algorithmic decision-making. Proposals include requiring platforms to publish detailed explanations of how their algorithms work, subjecting these systems to third-party audits, and allowing researchers controlled access to platform data for academic study. The European Union's Digital Services Act, for instance, includes provisions for algorithmic transparency and accountability, mandating that large platforms conduct risk assessments and share information with regulators about how their algorithms might amplify harmful content (European Commission, 2022).

In the European Union, the proposed Digital Services Act (DSA) and Digital Markets Act (DMA) aim to create a more standardized framework for algorithmic accountability by requiring large platforms to share information with regulators and undergo periodic audits. Although these legislative efforts are still in progress, they represent a significant step toward establishing clearer accountability standards.

In emerging markets like Bangladesh, however, the debate over algorithmic governance is often overshadowed by more immediate concerns about state censorship and disinformation. Regulatory frameworks in such contexts tend to focus on penalizing "offensive" content or prosecuting individuals for spreading false information, rather than addressing the opaque decision-making processes behind content curation. Consequently, while algorithmic transparency is a prominent topic in academic and policy circles in developed countries, it remains a secondary issue in nations where the regulatory focus is on controlling the message rather than on ensuring a fair information ecosystem.

4.3.3 Ethical Dilemmas and Platform Responses

Digital platforms have recognized the growing scrutiny over their algorithmic systems and have occasionally taken steps to address public concerns. For instance, Facebook has experimented with providing users more control over their news feed, allowing them to adjust their preferences for types of content. Twitter has introduced features that explain why certain tweets appear in a user's timeline, and YouTube has enhanced its content moderation policies to remove harmful or misleading videos. Despite these initiatives, critics argue that such measures are often cosmetic and fail to address the root causes of algorithmic bias. The profit motive inherent in the design of these systems where engagement metrics directly correlate with revenue creates a fundamental conflict of interest, as the most engaging content is not necessarily the most truthful or balanced (Zuboff, 2019).

Moreover, the ethical dilemmas extend beyond content curation. Platforms are increasingly experimenting with artificial intelligence to detect and remove harmful content. However, AI-driven moderation systems are prone to errors, particularly in languages and cultural contexts that are underrepresented in training datasets. In Bangladesh, for example, automated systems may struggle to accurately interpret nuances in regional dialects or detect satire in Bengali, leading either to over-censorship or to the unchecked spread of disinformation (Rahman & Tania, 2020). Such challenges underscore the need for locally adapted solutions that combine automated tools with human oversight.

In this context, the concept of "algorithmic accountability" becomes critical. It demands that companies not only disclose information about their algorithms but also subject them to independent evaluations. Some scholars advocate for the creation of an independent regulatory body that could oversee algorithmic practices and enforce compliance with human rights standards. Although such proposals are still in the realm of academic debate, they point to a future in which technology companies might be held to similar standards as traditional media organizations, ensuring that the benefits of personalization do not come at the expense of fairness and transparency.

4.4 Civil Liberties Concerns: Balancing National Security and Freedom of Speech

4.4.1 Surveillance Practices and the Expansion of State Power

One of the most contentious aspects of digital regulation is the tension between national security imperatives and the protection of civil liberties. In recent years, governments across the globe have expanded their surveillance capabilities in the name of combating terrorism, cybercrime, and other threats. In the United States, the post-9/11 era ushered in a series of measures including the USA PATRIOT Act and subsequent reauthorizations that granted federal agencies unprecedented access to digital communications (Zuboff, 2019). Revelations by Edward Snowden in 2013 exposed the extensive surveillance apparatus operated by the National Security Agency (NSA), igniting a fierce debate over the balance between security and privacy (Greenwald, 2014).

In contrast, emerging democracies such as Bangladesh have adopted more direct and often draconian measures. The Digital Security Act (DSA) of Bangladesh is emblematic of this approach. Under the DSA, law enforcement agencies have broad authority to intercept digital communications, monitor social media activity, and prosecute individuals for a range of "offenses" that are often vaguely defined (Human Rights Watch, 2021). While state authorities claim that such measures are necessary to maintain public order and prevent the spread of harmful propaganda, critics argue that they provide cover for political repression and stifle legitimate dissent.

The justification for state surveillance is frequently couched in terms of national security. Governments argue that in a world where terrorist networks and extremist groups operate online, robust surveillance is essential to preempt and disrupt potential threats. However, the scope and scale of surveillance raise profound questions about privacy and civil liberties. In many cases, surveillance programs lack adequate oversight, enabling abuse by security agencies. The risk of mission creep where surveillance tools initially deployed for counterterrorism are later used to monitor political opponents, activists, and journalists is particularly acute in countries with weak judicial independence.

The institutional consequences of expanded surveillance capabilities extend beyond privacy concerns. Law enforcement agencies, from the FBI to local police departments, have developed specialized units for social media monitoring, fundamentally changing how these institutions approach investigation and intelligence gathering. These units often operate with minimal judicial oversight, creating institutional power imbalances that favor executive agencies over judicial review. In Bangladesh, similar institutional shifts have occurred, with the Rapid Action Battalion (RAB) and Detective Branch establishing dedicated cyber monitoring cells that track political opposition activity on social media, often without warrants or formal judicial authorization (Human Rights Watch, 2021). This expansion of surveillance capacity has created new institutional structures within law enforcement that prioritize digital monitoring over traditional investigative methods, raising questions about the proper balance between security and liberty in the digital age.

4.4.2 *Censorship, Content Removal, and the Chilling Effect on Free Speech*

Beyond surveillance, censorship and content removal are other mechanisms by which states exert control over digital spaces. In the United States, where the First Amendment places strict limits on government censorship, direct content regulation is relatively rare. However, indirect pressure on platforms through congressional hearings, public shaming, or threats of regulatory action can achieve similar outcomes. For instance, during the 2020 U.S. presidential election, major social media platforms removed or flagged content related to election fraud claims, often in response to government and civil society pressure. While these actions were framed as necessary to combat misinformation, they also raised concerns about the blurring of lines between voluntary self-regulation and de facto state censorship.

In contrast, countries like Bangladesh employ more direct forms of censorship. The Digital Security Act empowers authorities to order the removal of content deemed defamatory, blasphemous, or threatening to national security. Critics point out that the law's vague language and lack of judicial oversight enable arbitrary enforcement, often targeting journalists, bloggers, and opposition activists (Article19, 2021). High-profile cases include the arrest of cartoonists for satirical depictions of government officials and the detention of students for critical social media posts. Such actions have a chilling effect on free speech, as individuals self-censor out of fear of legal repercussions.

The global nature of digital platforms complicates these censorship efforts. When a government demands the removal of content, it must often negotiate with multinational corporations that operate under different legal jurisdictions. In some cases, platforms comply with local laws to maintain access to lucrative markets, even when such laws conflict with international human rights standards. In other cases, they resist, leading to standoffs that can result in bans or shutdowns. For example, Turkey, Russia, and Pakistan have at various times blocked access to Twitter or YouTube in response to refusals to remove specific content, raising concerns about the fragmentation of the global internet and the erosion of digital freedoms (Freedom House, 2020).

4.4.3 *Privacy, Encryption, and the Debate Over Access*

Another dimension of the civil liberties debate centers on encryption and privacy. End-to-end encryption, a technology that secures communications so that only the sender and recipient can read them, has become a standard feature on many messaging apps, including WhatsApp, Signal, and Telegram. While encryption is celebrated for protecting user privacy and enabling secure communication, it also poses challenges for law enforcement. Governments argue that encryption creates "safe havens" for criminals and terrorists, making it difficult to intercept communications even with a warrant.

This tension came to a head in the United States during the 2016 San Bernardino case, when the FBI sought to compel Apple to create a backdoor into an iPhone used by the attackers. Apple refused, arguing that creating such a backdoor would undermine the security of all its devices and set a dangerous precedent (Greenberg, 2016). The case sparked a fierce debate about the balance between security and privacy, with technologists, civil libertarians, and law enforcement officials taking opposing sides.

In Bangladesh, the use of encryption has become a double-edged sword. Opposition activists and civil society groups rely on encrypted messaging apps to organize and communicate safely, especially in the face of government surveillance. However, the authorities view encryption with suspicion, occasionally pressuring platforms to provide access to encrypted communications. The lack of robust legal protections for privacy in Bangladesh means that users face significant risks, with little recourse if their communications are intercepted or misused (EFF, 2021).

4.5 Role of International Organizations and Global Watchdogs

As digital regulation becomes increasingly important, international organizations and global watchdogs have emerged as key players in shaping norms and standards. Organizations such as the United Nations, UNESCO, and Reporters Without Borders (RSF) have developed frameworks and guidelines aimed at protecting freedom of expression, privacy, and access to information in the digital age. These initiatives often emphasize the need for transparency, accountability, and respect for human rights, serving as benchmarks against which national policies can be evaluated.

UNESCO's Internet Universality Indicators, for example, provide a comprehensive framework for assessing whether national policies align with principles such as rights-based approaches, openness, accessibility, and multi-stakeholder participation (UNESCO, 2019). Similarly, RSF's World Press Freedom Index tracks violations of press freedom, including those related to digital censorship and surveillance, offering a comparative measure of how countries fare in protecting journalistic freedoms.

The Global Network Initiative (GNI), a multi-stakeholder organization that includes technology companies, civil society groups, and academics, has also played a role in promoting responsible business practices. The GNI's principles on freedom of expression and privacy set out guidelines for how companies should respond to government demands for user data or content removal, emphasizing the importance of transparency and respect for human rights (GNI, 2018). Although adherence to these principles is voluntary, they have influenced corporate policies and provided a framework for accountability.

Despite these efforts, the influence of international organizations is often limited by the sovereignty of nation-states and the economic interests of powerful actors. Countries with poor records on digital rights may simply ignore international guidelines, while multinational corporations may prioritize market access over human rights compliance. Moreover, the rapid pace of technological change means that international norms often lag behind emerging challenges, such as deepfakes, AI-driven surveillance, and new forms of online manipulation.

4.6 Toward a Convergent Future?

The regulatory landscape for digital media is characterized by significant divergence. On one end of the spectrum, the United States champions a model of minimal state intervention, relying on platform self-regulation and the protection of free speech. On the other end, countries like Bangladesh adopt more interventionist approaches, with broad powers for state authorities to censor and prosecute online content. Between these extremes lie numerous hybrid models, each reflecting unique political, economic, and cultural contexts.

Yet, as the digital landscape evolves, common challenges emerge that demand cooperative solutions. The rapid spread of disinformation, the opaque workings of algorithmic systems, and the tension between state security and individual liberties are not confined to any single nation; they are global phenomena that threaten democratic processes and social cohesion.

A convergent future for digital governance may lie in hybrid models that draw on the strengths of both centralized and decentralized approaches. In the United States, for instance, growing public concern over misinformation and political polarization has led some policymakers to advocate for incremental reforms to Section 230, alongside enhanced transparency in content moderation practices. In the European Union, regulatory experiments such as the DSA represent a middle ground balancing the need for corporate accountability with the preservation of free expression and market innovation. For emerging democracies like Bangladesh, reforms might focus on narrowing the scope of broadly defined laws like the DSA while simultaneously investing in digital literacy programs and independent oversight bodies that empower citizens rather than intimidate them.

Civil society plays an indispensable role in this evolving landscape. Grassroots organizations, media watchdogs, and academic institutions are increasingly mobilizing to educate the public about digital rights, advocate for transparent policies, and challenge overreaching censorship or surveillance. In Bangladesh, local NGOs have organized workshops to train journalists in digital safety and fact-checking, while international bodies like RSF and UNESCO work to hold governments accountable for repressive practices. At the same time, technology companies themselves are beginning to recognize that a sustainable business model requires ethical governance. The establishment of oversight boards, internal ethics committees, and collaborations with external auditors signal a growing awareness that the long-term health of the digital ecosystem depends on balancing profit with public interest.

The prospect of a truly convergent model of digital regulation is daunting but not unattainable. It will require ongoing dialogue among all stakeholders governments, corporations, civil society, and international organizations to create a dynamic regulatory environment that adapts to technological change without compromising fundamental rights. Multilateral initiatives, cross-border legal frameworks, and transnational research collaborations can help bridge divergent approaches, ensuring that digital governance evolves in a manner that protects both innovation and individual freedoms.

Ultimately, the future of digital regulation is likely to be an iterative process a series of adjustments and refinements rather than a single, definitive solution. As societies worldwide confront new challenges posed by artificial intelligence, deepfakes, and increasingly sophisticated cyber threats, the need for adaptive, resilient, and inclusive regulatory frameworks will only grow. The ongoing debates over censorship, algorithmic governance, and civil liberties are not merely technical policy discussions; they are fundamentally about the kind of societies we wish to build in the digital age. Balancing security with freedom, state oversight with private innovation, and local priorities with global norms is a complex endeavor, but it is essential for ensuring that the internet remains a tool for empowerment rather than repression.

4.7 State Regulation in South Asia India, Pakistan and Bangladesh

In South Asia, where democratic processes often converge with a mosaic of sociocultural and economic complexities, the regulation of digital media takes on unique significance. Bangladesh, India, and Pakistan each exhibit varying degrees of governmental oversight, legal codifications, and enforcement mechanisms aimed at mitigating online threats ranging from misinformation and hate speech to national security concerns. Yet these measures also raise persistent questions about free expression, civil liberties, and the increasingly blurred boundaries between private and public communication. This section offers a comparative, theoretical perspective on how three prominent South Asian states have grappled with the challenges of platform governance, censorship, and digital surveillance. In doing so, it highlights the tension between governments' perceived imperative to maintain societal order and critics' insistence on safeguarding foundational democratic norms an evolving tension that foreshadows the subsequent introduction of a more unified theoretical framework (D^3) in later chapters (see Habermas, 1989; McCombs & Shaw, 1972).

4.7.1 Regulatory Backdrop and Theoretical Underpinnings

From a theoretical standpoint, much of the existing literature on media governance in South Asia can be situated within a broader global discourse on media system dependency (Ball-Rokeach & DeFleur, 1976) and the interplay of state power with local socio-political contexts (Hallin & Mancini, 2004). While classical models often treat regulation as an external corrective mechanism imposed upon traditional mass media, the rise of networked platforms has complicated these assumptions (Norris, 2009; Pariser, 2011). Contemporary scholarship emphasizes that algorithmic amplification and user-driven content creation have effectively diminished traditional editorial gatekeeping, giving rise to misinformation and digital echo chambers (Banaji & Bhat, 2020; Sunstein, 2017). In states where democratic institutions are still consolidating, such as Bangladesh, India, and Pakistan, the regulatory instinct tends to be sharpened by concerns about extremist mobilization, ethnic and religious violence, and political partisanship (Ahmed, 2020; Zia, 2018). Consequently, governments often enact legislation that grants broad powers to law enforcement agencies even when this broadness may suppress legitimate dissent and curtail free speech in the name of public order.

4.7.2 India: Between Liability, Governance, and Oversight

India's digital regulation landscape is as vast and varied as its electorate. The Information Technology Act (IT Act) of 2000 has historically anchored digital governance, but evolving political and technological realities have spurred multiple amendments. Notably, Section 69A empowers the government to order the blocking of content that jeopardizes national security or public order (Banaji & Bhat, 2020). In 2021, the government introduced the Information Technology (Intermediary Guidelines and Digital Media Ethics Code) Rules, intensifying obligations for platforms to remove "unlawful" content promptly and requiring them to comply with data and traceability directives (Chowdhury, 2022). These measures echo broader theoretical insights on computational propaganda (Woolley & Howard, 2018), as Indian authorities cite the proliferation of online rumors especially on WhatsApp as justification for robust takedown protocols.

However, such regulation also illustrates an algorithmic governance dilemma (Pasquale, 2015). While these measures aim to manage extremist or misleading content, they rely on technology companies whose underlying recommendation systems may unintentionally amplify sensational or polarizing messages (Pariser, 2011). The tension between ensuring public safety and preserving digital freedoms is acute in India's multi-ethnic, multi-religious context, where communal grievances can escalate quickly. Critics warn that expansive regulatory powers can be wielded selectively, targeting dissent under the guise of preventing hate speech. This selective enforcement resonates with theories of agenda-setting and priming (McCombs &

Shaw, 1972), as certain narratives gain governmental endorsement while others are suppressed.

4.7.3 Pakistan: Security Paradigms and the Digital Sphere

Pakistan's approach to digital regulation is heavily influenced by a security-centric paradigm, shaped by ongoing concerns about terrorism, separatism, and external interference. The Prevention of Electronic Crimes Act (PECA) of 2016 grants authorities broad powers to monitor online communications, block websites, and prosecute individuals for cyber offenses (Zia, 2018). While the act targets cybercrimes such as hacking, identity theft, and online harassment, it also contains provisions for regulating "hate speech" and content that undermines state sovereignty language that has been criticized as vague and susceptible to misuse.

Like India, Pakistan has witnessed repeated clashes between the state and digital platforms over content takedown requests. In 2020, for instance, the government temporarily blocked access to TikTok, citing concerns over "immoral" and "indecent" content (Hussain & Khan, 2021). Critics contend that such bans reflect a broader pattern of using national security and moral policing as pretexts for controlling online discourse. Scholarly explorations of computational propaganda (Woolley & Howard, 2018) also suggest that troll farms and bot networks operate in tandem with official narratives, further entangling state structures and digital ecosystems. This dual use of regulation both to control extremist speech and to constrain broader political critique exemplifies the complex interplay of political power and media technologies (Benkler, 2006).

4.7.4 Bangladesh: Digital Security and Expanding State Powers

Bangladesh's journey in digital governance illustrates how emerging democracies negotiate the twin pressures of rapid technological expansion and persistent political volatility (Islam & Sultana, 2021). The Digital Security Act (DSA), introduced in 2018, quickly became a linchpin of debate. Officially intended to address cybercrime, religious incitement, and so-called "anti-state" activities, the DSA grants law enforcement the authority to arrest individuals without a warrant under loosely defined categories such as "hurting religious sentiment" or "propagating rumors" (Ahmed, 2020). Critics argue that these provisions align more with the Polarized Pluralist Model where media often aligns with political or ideological factions (Hallin & Mancini, 2004) as state agencies selectively enforce regulations against opposition supporters or journalists. Indeed, multiple reports by Reporters Without Borders (2022) have documented how bloggers, activists, and even cartoonists have faced arrest or intimidation for digital content deemed defamatory toward ruling-party figures.

Theoretically, Bangladesh's approach highlights what might be termed "hyper-dependency on regulatory intervention," in which the state faced with historically volatile political contests views online platforms as catalysts for potential unrest (Rahman & Chowdhury, 2022). This climate often stokes self-censorship among citizens who fear retribution under broadly worded legal statutes (Ahmed, 2020). From a broader lens, it underscores the limitations of classical agenda-setting theory (McCombs & Shaw, 1972) to explain regulation in digital contexts; rather than a singular "agenda," multiple state and non-state actors political parties, civil society, diaspora communities vie to shape public opinion, leading the government to enact wide-ranging laws to reclaim narrative dominance. This tension paves the way for more holistic theoretical frameworks such as the forthcoming D^3 model which can account for the interplay of digital fragmentation, disinformation flows, and asymmetrical state power.

The DSA's impact on institutional behavior extends beyond individual self-censorship to organizational caution. Newsrooms have established internal review processes for social media posts, effectively creating additional editorial layers that slow journalism and reduce its responsiveness. Universities have issued guidelines discouraging faculty from political commentary on social media, constraining academic freedom. Even government officials in lower bureaucratic positions express reluctance to engage on social media for fear that old posts might be weaponized against them, reducing government transparency and public engagement (KII #8, 2025). These institutional adaptations reveal how digital regulation shapes not just individual behavior but the operational norms of organizations across civil society, media, and government, creating a pervasive culture of caution that fundamentally alters how institutions engage with the digital public sphere.

4.8 Convergence and Divergence in South Asian Regulatory Models

Although Bangladesh, India, and Pakistan differ in historical trajectory, economic capacity, and political culture, they converge in deploying broad regulatory tools as a means of managing contentious digital spaces. Each nation justifies its legal framework through reference to national security, public order, or communal harmony, yet the practical implementation often reveals deep-seated anxieties about losing control over public discourse. This shared pattern suggests that South Asian democracies perceive digital platforms as double-edged swords: potent means for citizen empowerment and political mobilization, yet equally capable of inflaming existing social fissures or undermining state authority (Benkler, 2006; Tufekci, 2017).

At the same time, divergences remain significant. India, with its robust judiciary and relatively independent press, has witnessed greater public debate over regulatory overreach, even as the government pushes for expansive content regulation. Pakistan's military-influenced governance shapes how digital laws are enacted and enforced, with security agencies playing a prominent role. Bangladesh's DSA stands out for its sweeping powers to prosecute individuals without judicial warrants, reflecting both a fragile democratic environment and an executive branch emboldened to suppress dissent through legal means. Such differences underscore that while technology may be universal, the political and social contexts into which it is integrated remain deeply particular.

4.8.1 Institutional Vulnerabilities Across Contexts

Despite differing regulatory approaches, democratic institutions in both developed and developing contexts face common vulnerabilities to social media influence:

1. Resource Asymmetry: Government institutions typically lack the technical resources and expertise of the platforms they attempt to regulate, creating power imbalances that favor corporate actors over public accountability. Electoral commissions cannot match the data analytics capabilities of Facebook or Google, courts struggle to evaluate algorithmic evidence without forensic expertise, and legislatures lack the technical staff to conduct meaningful oversight of platform operations.
2. Speed Mismatch: Institutional decision-making processes designed for deliberation cannot keep pace with real-time social media dynamics, allowing misinformation to spread and solidify before institutional responses mobilize. By the time electoral commissions investigate complaints of digital manipulation, elections have concluded; by the time courts rule on content disputes, viral posts have already shaped public opinion; by the time legislatures debate platform regulation, new technologies have emerged.
3. Jurisdictional Confusion: Social media's borderless nature creates jurisdictional questions that traditional institutional boundaries cannot accommodate, leading to enforcement gaps and regulatory arbitrage. When content crosses borders, which nation's laws apply? When platforms are headquartered abroad, how can local institutions compel compliance? When diaspora communities coordinate political action from overseas, which government has jurisdiction?
4. Accountability Diffusion: When responsibility for digital governance is distributed across multiple institutions (legislatures, courts, regulatory agencies, platforms themselves), accountability becomes difficult to assign and enforce. If misinformation spreads during an election, is the platform responsible for not removing it, the electoral commission for not monitoring it, the court for not prosecuting it, or the legislature for not regulating it adequately? This diffusion of responsibility creates institutional paralysis where no single entity feels accountable for addressing systemic problems.

These common vulnerabilities suggest that institutional reform must accompany regulatory innovation if democratic systems are to effectively govern social media's influence on political life. Institutions need not just new laws but new capacities: technical expertise, rapid response mechanisms, cross-border cooperation frameworks, and clear accountability structures. Without such institutional strengthening, even well-designed regulations will fail in implementation (Table 4.1).

Key Insights from Comparative Analysis

Institutional Convergence: Despite vastly different political systems and regulatory philosophies, both U.S. and Bangladeshi institutions face similar structural challenges: knowledge gaps, resource constraints, speed mismatches, and accountability diffusion. These challenges transcend political context and represent fundamental incompatibilities between traditional institutional structures and digital communication dynamics.

Divergent Responses: While challenges converge, institutional responses diverge based on political context. U.S. institutions tend toward slow, deliberative adaptation within existing frameworks, constrained by constitutional protections and political polarization. Bangladeshi institutions adopt more interventionist approaches with greater political discretion, enabled by weaker judicial independence and executive dominance. These divergent responses reflect deeper differences in democratic culture and state capacity.

Democratic Health Indicators: The health of democratic institutions correlates with their capacity to adapt to social media's influence without sacrificing core principles (independence, transparency, accountability). Both contexts show institutional strain, though manifested differently. U.S. institutions maintain greater independence but struggle with gridlock and capture by political factions; Bangladeshi institutions show more decisiveness but lack independence and accountability. Neither model successfully balances adaptation with principle preservation.

Implications for Institutional Reform: Effective institutional response to social media requires not just regulatory reform but structural transformation. Institutions need technical capacity building, rapid response mechanisms that preserve deliberation, cross-border cooperation frameworks, and clear accountability structures. Without such reforms, even well-designed regulations will fail in implementation. The comparative analysis suggests that institutional strengthening may be more critical than specific regulatory approaches for democratic resilience in the digital age.

4.9 Conclusion

The evolution of digital regulatory frameworks in the twenty-first century reflects a profound transformation in how societies manage information, security, and individual rights. In the United States, a model of self-regulation backed by constitutional free speech protections has fostered a vibrant, albeit sometimes chaotic,

Table 4.1 Institutional influence of social media across democratic contexts

Institution	United States	Bangladesh	Common patterns
Electoral Management	FEC struggles with digital ad disclosure; limited resources for monitoring; jurisdictional gaps	EC establishes monitoring units but lacks capacity; manual reporting systems; limited independence	Institutional lag behind technology; resource constraints limit effectiveness
Judicial System	Section 230 cases show inconsistent interpretation; courts develop digital evidence protocols slowly	DSA cases overwhelm courts; judges lack digital literacy training; screenshot evidence controversies	Institutional knowledge gaps; rapid case volume growth; outdated legal frameworks
Legislative Bodies	Congressional hearings become social media performances; individual member branding over party positions	Parliament debates risk DSA prosecution when shared online; MPs use Facebook to bypass traditional media	Social media pressure shapes agendas; institutional coherence weakened; individual over collective voice
Traditional Media	Newspaper collapse; TV news chases social media trends; advertising revenue decimated	Print struggles; TV legitimizes unverified social content; investigative capacity declines	Economic disruption; content sourcing from social media; reduced accountability journalism
Civil Society	NGOs master digital advocacy but face astroturfing and coordinated attacks	Organizations document abuses via social media while facing government surveillance	Digital capacity gaps; surveillance vulnerabilities; coordination benefits vs. security risks
Political Parties	Digital operations decentralize party power; outsider candidates bypass traditional structures	Government parties establish cyber teams; opposition relies on diaspora networks	Message discipline challenges; hierarchical to networked structures; resource-dependent capacity
Law Enforcement	FBI and local police develop social media monitoring units with minimal oversight	RAB and Detective Branch create cyber cells for political monitoring without warrants	Expanded surveillance capacity; reduced judicial oversight; institutional power imbalances
Regulatory Agencies	Agencies lack technical expertise to match platform sophistication; reactive rather than proactive	Limited regulatory infrastructure; selective enforcement favoring government interests	Technical knowledge gaps; resource asymmetry with platforms; enforcement inconsistency

digital marketplace. Meanwhile, in countries like Bangladesh, direct state intervention through measures such as the Digital Security Act has become the preferred approach, albeit one that raises serious concerns about censorship and the stifling of dissent.

At the heart of these debates lies the challenge of algorithmic governance. As algorithms increasingly shape public discourse by determining what information is seen and by whom the need for transparency, accountability, and independent oversight has never been more pressing. Equally critical is the balancing act between national security and civil liberties. Whether through surveillance programs in the United States or broad content removal powers in Bangladesh, the tension between protecting citizens and preserving their freedoms remains a central and unresolved dilemma.

Beyond regulatory frameworks and algorithmic governance, this chapter has revealed profound institutional transformations driven by social media's influence on democratic systems. Electoral bodies, courts, legislatures, media organizations, civil society, and political parties the foundational institutions of democratic life face common challenges of adaptation, resource constraints, and accountability diffusion. While regulatory approaches differ significantly between the United States' self-regulation model and Bangladesh's interventionist framework, the institutional impacts show remarkable convergence: knowledge gaps, speed mismatches with digital dynamics, and power asymmetries favoring platforms over public institutions.

The comparative analysis reveals that institutional vulnerabilities may pose greater threats to democratic resilience than specific regulatory choices. When electoral commissions lack resources to monitor digital campaigns, when courts cannot evaluate algorithmic evidence, when legislatures respond to social media pressure rather than deliberative processes, when traditional media collapses under economic disruption, when civil society faces surveillance threats, and when political parties fragment under digital decentralization, the cumulative effect undermines democratic governance regardless of regulatory philosophy.

International organizations and global watchdogs offer a glimmer of hope by establishing shared standards and fostering dialogue among disparate regulatory systems. Initiatives such as UNESCO's Internet Universality Indicators and RSF's press freedom rankings provide benchmarks that can guide national reforms and encourage cross-border cooperation. Although the concept of a "Digital Geneva Convention" remains aspirational, it signals a growing recognition that global challenges require global solutions.

Looking ahead, the path to a convergent future in digital regulation will likely involve hybrid models that draw on the strengths of both decentralized and centralized approaches. It will require a sustained commitment to transparency, ethical governance, and the protection of fundamental rights, as well as ongoing dialogue among governments, corporations, and civil society. Critically, it will require institutional innovation alongside regulatory reform. Democratic institutions must develop technical capacity, accelerate decision-making processes without sacrificing deliberation, establish clearer accountability mechanisms, and build resilience against both disinformation and overreach. The future of digital democracy depends less on any single regulatory framework than on the adaptive capacity of democratic institutions themselves.

While no single framework can perfectly resolve the inherent tensions between security and freedom, continued innovation and cooperation offer the best hope for a digital future that empowers individuals, fosters democratic debate, and upholds the values of an open society. The path forward requires recognizing that regulatory frameworks operate through institutions, and institutional health determines regulatory effectiveness. As societies confront emerging challenges from artificial intelligence, deepfakes, and sophisticated cyber threats, strengthening democratic institutions not just crafting better regulations must become the priority for safeguarding democracy in the digital age.

In sum, the regulatory frameworks shaping our digital world are at a critical juncture. The choices made today about censorship, algorithmic transparency, and state surveillance will have profound implications for the nature of public discourse, the resilience of democratic institutions, and the rights of individuals in the years to come.

Case Study

The contrast between U.S. Section 230 and Bangladesh's Digital Security Act (DSA) illustrates divergent regulatory philosophies. Section 230 shields platforms from liability for user content, enabling innovation but also permitting harmful speech to proliferate. The DSA, conversely, imposes criminal penalties for "propaganda" and content that "tarnishes" state image—ostensibly targeting misinformation but weaponized against journalists and opposition voices.

Estonia's 2025 AI-assisted governance pilot demonstrates a third approach: algorithmic transparency combined with human oversight. The European Union's Digital Services Act attempts to mandate content moderation standards while preserving free expression. Each framework reflects different balances between innovation, security, and civil liberties, shaped by distinct political cultures and democratic traditions.

Questions for Discussion

1. How can societies balance platform accountability with free speech protections in regulating social media?
2. What mechanisms ensure that anti-misinformation laws do not become tools for authoritarian censorship?
3. Should algorithmic decision-making in content moderation be subject to mandatory transparency requirements?
4. How do surveillance practices enabled by digital regulation affect civil liberties and political dissent?
5. What role should international organizations play in establishing global norms for digital governance?

Food for Thought

When private corporations control public discourse infrastructure, who should set the rules—democratic governments, market forces, or international bodies—and what legitimizes that authority?

What We Learn

Regulatory frameworks reveal fundamental tensions between state sovereignty and global interconnectedness, between security and liberty, between innovation and accountability. Effective regulation requires participatory processes that include civil society, technical experts, and affected communities—not just governments and platforms. The goal is not perfect control but adaptive governance that can evolve with technology while safeguarding democratic principles.

References

Ahmed, F. (2020). Troll armies in emerging democracies: Bangladesh's online partisans. *Journal of Political Communication, 15*(3), 210–228.

Article 19. (2021). *Blocking, filtering and takedown of content: The legal framework*. Retrieved from https://www.article19.org

Ball-Rokeach, S. J., & DeFleur, M. L. (1976). A dependency model of mass-media effects. *Communication Research, 3*(1), 3–21.

Banaji, S., & Bhat, R. (2020). WhatsApp vigilantes: An exploration of citizen reception and circulation of WhatsApp misinformation linked to mob violence in India. In *London School of Economics and Political Science.*

Benkler, Y. (2006). *The wealth of networks: How social production transforms markets and freedom.* Yale University Press.

Benkler, Y., Faris, R., & Roberts, H. (2018). *Network propaganda: Manipulation, disinformation, and radicalization in American politics.* Oxford University Press.

Chowdhury, A. R. (2022). India's new intermediary guidelines: Implications for platform governance and digital rights. *South Asian Journal of Law and Policy, 8*(2), 45–68.

EFF (Electronic Frontier Foundation). (2021). *EFF's work on encryption.* Retrieved from https://www.eff.org

European Commission. (2022). *The digital services act package.* Retrieved from https://ec.europa.eu/digital-services-act

Freedom House. (2020). *Freedom on the net 2020: The pandemic's digital shadow.* Retrieved from https://freedomhouse.org

GNI (Global Network Initiative). (2018). *Principles on freedom of expression and privacy.* Retrieved from https://globalnetworkinitiative.org

Goldman, E. (2020). *An overview of Section 230.* Santa Clara University School of Law.

Greenberg, A. (2016). Apple versus FBI: The battle to unlock an iPhone and the end of encryption. *Wired.* Retrieved from https://www.wired.com

Greenwald, G. (2014). *No place to hide: Edward Snowden, the NSA, and the U.S. surveillance state.* Metropolitan Books.

Habermas, J. (1989). *The structural transformation of the public sphere: An inquiry into a category of bourgeois society.* MIT Press.

Hallin, D. C., & Mancini, P. (2004). *Comparing media systems: Three models of media and politics.* Cambridge University Press.

Human Rights Watch. (2021). *Muzzling dissent: The impact of Bangladesh's digital security act*. Retrieved from https://www.hrw.org

Hussain, M., & Khan, A. (2021). Digital censorship in Pakistan: The TikTok ban and moral policing online. *Media Asia, 48*(3), 178–194.

Islam, M. S., & Sultana, R. (2021). Digital governance in emerging democracies: Bangladesh's regulatory challenges. *Asian Journal of Political Science, 29*(1), 89–107.

Kosseff, J. (2019). *The twenty-six words that created the internet*. Cornell University Press.

Kreiss, D. (2016). *Prototype politics: Technology-intensive campaigning and the data of democracy*. Oxford University Press.

Kreiss, D., & McGregor, S. C. (2019). Technology firms shape political communication. *Political Communication, 36*(3), 264–266.

McCombs, M. E., & Shaw, D. L. (1972). The agenda-setting function of mass media. *Public Opinion Quarterly, 36*(2), 176–187.

Norris, P. (2009). *Driving democracy: Do power-sharing institutions work?* Cambridge University Press.

O'Donovan, C., Warzel, C., & McDonald, L. (2019). YouTube's recommendation algorithm has a dark side. *BuzzFeed News*. Retrieved from https://www.buzzfeednews.com

Pariser, E. (2011). *The filter bubble: How the new personalized web is changing what we read and how we think*. Penguin.

Pasquale, F. (2015). *The black box society: The secret algorithms that control money and information*. Harvard University Press.

Pew Research Center. (2021). News use across social media platforms in 2021. Retrieved from https://www.pewresearch.org

Rahman, T., & Chowdhury, M. A. (2022). Fake news and political misinformation in Bangladesh. *Journal of Media and Politics, 10*(3), 50–72.

Rahman, T., & Tania, T. (2020). Rural connectivity and digital divide: A comparative analysis of North and South Bangladesh. *Telecom Policy Studies, 12*(2), 56–70.

Reporters Without Borders. (2022). *2022 World Press freedom index*. Retrieved from https://rsf.org

Sunstein, C. R. (2017). *#Republic: Divided democracy in the age of social media*. Princeton University Press.

Tufekci, Z. (2017). *Twitter and tear gas: The power and fragility of networked protest*. Yale University Press.

UNESCO. (2019). *Internet universality indicators: A framework for assessing internet development*. Retrieved from https://en.unesco.org

Vaidhyanathan, S. (2018). *Antisocial media: How Facebook disconnects us and undermines democracy*. Oxford University Press.

Woolley, S. C., & Howard, P. N. (Eds.). (2018). *Computational propaganda: Political parties, politicians, and political manipulation on social media*. Oxford University Press.

Zia, A. S. (2018). Digital activism and state surveillance in Pakistan: The PECA challenge. *Journal of Asian and African Studies, 53*(6), 890–906.

Zuboff, S. (2019). *The age of surveillance capitalism: The fight for a human future at the new frontier of power*. PublicAffairs.

Chapter 5
Digital Media in the United States Cambridge Analytica, Data Harvesting, and the Politics of Persuasion

5.1 Introduction

Over the last two decades, social media has radically transformed how political actors communicate with the public in the United States. Platforms such as Facebook, Twitter (now X), and YouTube quickly evolved from innovative tech start-ups to colossal information ecosystems where election campaigns unfold in real-time. As online spaces gained prominence in civic life, they also created fertile ground for new forms of voter outreach, such as micro-targeting and data-driven persuasion strategies. This chapter examines the impact of digital media on the U.S. electoral landscape, centering on the Cambridge Analytica scandal, controversies over data harvesting, and debates around the 2016 and 2020 U.S. presidential elections.

We begin by charting the rise of social media giants and the emergence of micro-targeted political campaigns, demonstrating how platforms monetized user data to shape voter behavior. We then explore the details of Cambridge Analytica one of the most consequential scandals that exposed weaknesses in Facebook's data-sharing policies and how this episode underscored broader ethical dilemmas in digital campaigning. The chapter proceeds to the post-2016 landscape, detailing congressional hearings, public outcries, and evolving platform policies. Finally, we delve into ongoing debates about algorithmic accountability, the role of Section 230 of the Communications Decency Act, and how the digital sphere influenced both the 2016 and 2020 U.S. elections, including issues of Russian interference and post-election misinformation. Ultimately, these discussions illuminate the shifting power dynamics between tech platforms, political campaigns, and the American electorate in the digital age.

B. Ahmed and R. J. Samir, *Beyond Ballot Boxes*,
https://doi.org/10.1007/978-3-032-27350-5_5

5.2 The Rise of Social Media Giants and the Advent of Micro-targeting

5.2.1 Expansion of Platforms and User Engagement

In the early 2000s, platforms like Facebook (launched in 2004), YouTube (2005), and Twitter (2006) were novel tools that quickly became integral to social, economic, and political life. By the 2010s, Facebook boasted billions of users worldwide, with a significant share residing in the United States (Statista, 2022). Twitter, though smaller by user count, emerged as a hub for breaking news, political commentary, and real-time public discourse (Murthy, 2018). Meanwhile, YouTube transformed the landscape of video consumption, allowing political messages once confined to television to reach vast online audiences without the high costs of traditional broadcasting.

These platforms capitalized on a simple but powerful concept: user-generated content. Instead of relying on professional gatekeepers, social media empowered individuals to create, post, and share text, images, and videos. The ensuing network effects led to exponential growth. Increasingly, politicians saw social media as a more direct channel to reach voters, bypassing traditional media filters. Over time, these companies introduced tools such as Pages (for organizations and public figures), sponsored posts, and targeted advertising, laying the groundwork for sophisticated campaign strategies.

5.2.2 Emergence of Data-Driven Campaigning

While television and radio remain crucial for broad outreach, micro-targeting on social media allows political campaigns to deliver highly personalized messages. Relying on vast datasets that include demographic details (age, gender, location), behavioral metrics (likes, follows, click patterns), and even psychographic indicators (values, attitudes, personality traits), campaigns can tailor content to specific groups or individuals (Kreiss, 2016). This approach represents a leap from the "one-size-fits-all" messaging of earlier eras.

Granular Segmentation: Campaigns can, for instance, identify a suburban mother passionate about education reform and direct her to ads highlighting a candidate's stance on school funding. A young professional concerned about student debt might see tailored content on debt-relief policies.

Echo Chambers: As micro-targeting personalizes feeds, individuals may be less likely to encounter divergent viewpoints (Pariser, 2011). In theory, voters become locked in echo chambers, reinforcing pre-existing beliefs and fostering political polarization (Sunstein, 2017).

5.2.3 Regulatory Blind Spots

From a regulatory standpoint, the meteoric rise of social media caught legislators off guard. Campaign finance laws, written primarily for television and print media, did not fully address the complexities of digital advertising (Anstead, 2017). Regulations around disclaimers such as "paid for by" notices were inconsistent across platforms, and oversight agencies struggled to track and verify the origin or funding of ads, including those produced by foreign entities. This regulatory ambiguity would prove consequential in the run-up to the 2016 U.S. election.

5.3 Cambridge Analytica and the 2016 Election

Cambridge Analytica, formed in 2013, was affiliated with the British firm SCL Group. It specialized in data analytics to influence voter behavior through a combination of demographics, psychographics, and behavioral targeting (Grassegger & Krogerus, 2017). The firm's psychographic profiling categorized people based on traits like openness, conscientiousness, and neuroticism purporting to predict and manipulate political choices more accurately. Initially, Cambridge Analytica offered its services to various conservative campaigns in the U.S., including Ted Cruz's 2016 presidential run, before working with the Trump campaign.

5.3.1 The Data Harvesting Scandal

In March 2018, news outlets such as *The Guardian* and *The New York Times* revealed that Cambridge Analytica had illicitly acquired *personal data from over 50 million Facebook users* later estimates rose to 87 million. This breach stemmed from a third-party app called "This Is Your Digital Life," developed by a researcher who used Facebook's then-lax data-sharing policies. The app collected personal data from users and their network of friends, even if the friends never explicitly consented (Cadwalladr & Graham-Harrison, 2018).

Cambridge Analytica allegedly leveraged these data to build detailed voter profiles, enabling campaigns to bombard users with highly targeted political ads. Reports suggested that certain ads played on emotional triggers such as fear, nationalism, or resentment toward establishment politics. Critics claimed that such tactics transcended traditional campaigning, veering into *manipulative* or even *exploitative* territory (Benkler et al., 2018).

5.3.2 Ethical and Democratic Implications

Privacy Violations: Facebook users had not knowingly given consent to share their data with a political analytics firm. This raised alarms about *informed consent* and *user autonomy* in digital spaces.

Manipulation versus Persuasion: While political advertising has always aimed to influence voters, micro-targeting based on psychological vulnerabilities brought concerns about crossing ethical red lines.

Corporate Responsibility: Facebook faced scrutiny for failing to protect user data, prompting discussions about whether social media companies should serve as mere "neutral platforms" or take on more active responsibility for content governance.

These revelations sparked a global debate on *data privacy* and forced governments and citizens worldwide to confront the political ramifications of opaque data practices. While some argued that Cambridge Analytica's role in shaping the 2016 election might be overstated, the scandal underscored how insufficient oversight and accountability could jeopardize the integrity of democratic processes (Gelman & Azari, 2017).

5.3.3 Post-2016 Landscape: Congressional Hearings, Public Outcry, and Platform Policy Changes

In the wake of Cambridge Analytica, *Congressional hearings* became a focal point for lawmakers seeking answers from tech executives. Notably, *Mark Zuckerberg* (CEO of Facebook) testified before the Senate in April 2018, facing questions about user data protections, third-party applications, and the platform's role in disseminating misinformation (Tufekci, 2018). Lawmakers from both parties expressed alarm over foreign interference and the weaponization of social media platforms to sow political discord.

Disclosure Gaps: Investigators discovered that Facebook's advertising tools lacked robust measures to verify advertisers' identities. Consequently, foreign actors could purchase political ads targeting American voters with minimal scrutiny (House Permanent Select Committee on Intelligence, 2017).

Calls for Reform: Several bipartisan bills proposed new standards for digital political advertising, including mandatory disclaimers and stricter oversight to match or exceed those regulating TV ads (Kosseff, 2019).

5.3.4 Public Outcry and Grassroots Activism

News of data misuse, disinformation, and election meddling triggered a surge in *public backlash*. Activists led campaigns such as *#DeleteFacebook*, encouraging users to close their accounts. Although the movement did not result in a mass exodus from social platforms, polls indicated that trust in social media companies especially Facebook tumbled (Edelman Trust Barometer, 2019). Civil society organizations increasingly invested in *digital literacy* programs to help citizens recognize manipulated content and understand the implications of targeted ads. Fact-checking alliances proliferated to combat the spread of "fake news," emphasizing the shared responsibility among users, media outlets, and platforms to uphold factual discourse (Allcott & Gentzkow, 2017).

5.3.5 Platform Responses and Policy Revisions

Under intense scrutiny, tech giants announced measures aimed at enhancing *transparency* and *user control*:

Political Ad Archives: Facebook launched an online database displaying who purchased political ads, how much they spent, and which demographics they targeted (Facebook, 2018).

Stricter Developer Policies: Platforms revised application programming interfaces (APIs), restricting third-party data access. Apps seeking sensitive user data had to comply with new audit processes.

Content Moderation Overhauls: Twitter introduced new guidelines for political and issue-based ads and later banned most political advertising outright in 2019, citing concerns over potential misinformation (Twitter, 2019). YouTube enhanced its content moderation and labeling of election-related videos, though inconsistencies in enforcement remained (Lewis, 2020).

Even as these changes took effect, critics questioned their efficacy. Digital advertising remained easy to manipulate, and social networks continued to grapple with the volume of disinformation, trolling, and hateful content. Moreover, as new controversies arose, it became apparent that platform self-regulation might not suffice to address deeper systemic problems (Vaidhyanathan, 2018).

5.4 Ongoing Debates: Algorithmic Accountability, Section 230, and Big Tech Regulation

A significant revelation from the Cambridge Analytica episode was the *centrality of algorithms* in shaping user experiences on social media. Platforms rely on algorithms to prioritize certain content often sensational or emotionally charged because it drives engagement and advertising revenue (Benkler et al., 2018). Critics argue these algorithms foster political polarization and filter bubbles, as users see information that reinforces their existing biases (Pariser, 2011).

Calls for *algorithmic accountability* focus on demands that tech companies disclose how their ranking and recommendation systems work. Transparency advocates believe independent audits could detect *ideological skew* or *hidden biases* that might distort public debate. However, platform owners contend that revealing too much about these processes could enable malicious actors to exploit them, not to mention risking proprietary business secrets (Pasquale, 2015).

5.4.1 *Section* 230 and Liability Debatès

Section 230 of the Communications Decency Act (1996) is a core legal provision that immunizes digital platforms from liability for user-generated content. The law intended to promote free speech online by preventing a flood of litigation against internet service providers. Yet, in the wake of disinformation campaigns and extremist content, critics argue that Section 230 allows platforms to avoid responsibility for the societal harms they facilitate (Kosseff, 2019).

Proposed Reforms: Ideas range from narrowing Section 230 protections to removing them entirely for platforms that fail to demonstrate "good-faith" moderation.

Freedom of Speech Concerns: Tech companies, civil libertarians, and some lawmakers warn that repealing or weakening Section 230 might stifle open discourse, prompting overzealous censorship to avoid liability (Electronic Frontier Foundation, 2020).

The debate reflects a broader tension between *regulating* digital spaces to protect democratic norms and preserving the internet's open, innovative ethos. With diverse viewpoints across political lines, forging consensus on how to reform Section 230 remains challenging.

5.4.2 *From Self-regulation to Co-regulation?*

Some policymakers and researchers advocate for *co-regulation* a model where the government sets clear guidelines for content moderation, transparency, and data handling, but the platforms themselves handle day-to-day governance. The European Union has experimented with this approach through measures like the *General Data Protection Regulation (GDPR)* and the proposed *Digital Services Act*, influencing global conversations on privacy and platform responsibilities (Gorwa, 2019). In the United States, however, regulatory inertia and political polarization hamper broad legislative changes. The question remains whether rising public discontent will ultimately push Congress to take more aggressive steps or if incremental platform-led reforms will continue to define the U.S. model.

5.5 Case Highlights: 2016 and 2020 Elections

While *Cambridge Analytica* was central to discussions of data misuse, the *2016 election* also exposed vulnerabilities exploited by *foreign interference.* Investigations by the *House Permanent Select Committee on Intelligence* and the *Senate Intelligence Committee* revealed that Russian actors, particularly the *Internet Research Agency (IRA)*, orchestrated troll farms that created divisive social media content targeted at American users (DiResta et al., 2018). These posts spanned ideological spectrums, designed to polarize discussions on race, immigration, and religious issues. Fake pages posing as grassroots political groups engaged millions of U.S. voters.

Researchers suggest that while it is difficult to quantify the precise electoral impact of such interference, the *undermining of trust* in the electoral process and the fomenting of distrust among opposing political camps constituted a significant threat to American democracy. This episode underscored how unregulated social platforms could be weaponized by foreign actors with relative ease, prompting urgent calls for better ad verification and improved detection of coordinated inauthentic behavior (Bradshaw & Howard, 2019).

5.5.1 The 2020 Election, Post-election Controversies, and Intensified Misinformation

If the 2016 election sounded an alarm about digital vulnerabilities, the *2020 election* demonstrated that while platforms made some improvements (fact-checking labels, removal of coordinated disinformation networks), misinformation persisted at extraordinary levels. *Twitter* tagged and limited the reach of numerous tweets from high-profile political figures for containing false election-related claims. *Facebook* took steps to remove networks propagating QAnon conspiracy theories and more aggressively flagged unreliable content (Roose, 2021). However, critics argued these measures were often too little, too late.

The aftermath of the 2020 election saw a surge in *baseless allegations of voter fraud*, culminating in the *January 6th Capitol riot* an event partially organized through online forums and social media groups (Zolan, 2021). As tension escalated, companies faced pressure to de-platform individuals deemed responsible for inciting violence or spreading dangerous misinformation. *Twitter* permanently suspended President Donald Trump's account, while Facebook initially imposed an indefinite ban. These actions ignited new debates on *corporate censorship* and *platform power* raising questions about how private companies balance the public's interest in open debate with the imperative to prevent real-world harm.

5.5.2 *Donald Trump's 2024 Digital Campaign: Elon Musk, Truth Social, and the Legacy of Cambridge Analytica*

Context and return to social media

Following the intense controversies of the 2016 and 2020 campaigns both marked by micro-targeting scandals, data breaches, and platform interventions Donald Trump's bid for the 2024 U.S. presidency illustrated how drastically the digital environment had continued to evolve. While previous chapters outlined the Cambridge Analytica revelations and subsequent regulatory debates, the 2024 cycle demonstrated that partial legislative fixes and platform policy tweaks had not extinguished the powerful incentives fueling inflammatory and viral messaging. When Elon Musk acquired Twitter in late 2022, he enacted sweeping changes to content moderation, describing his vision as "maximal free speech." This pivot allowed numerous suspended accounts to be reinstated, including Trump's, whose ban had been emblematic of the post-January 6th clampdown on extremism (Twitter, 2023). Almost overnight, a candidate who had once relied heavily on Twitter to shape media narratives found himself free to broadcast in real time. Yet this was only half the story. Over the previous year, Trump had also cultivated his own platform, *Truth Social* an ideologically insulated microblogging site that functioned as a "safe harbor" from fact-checking disclaimers and mainstream moderation. The synergy between Musk's newly permissive Twitter and the quasi-closed ecosystem of Truth Social proved a strategic linchpin in Trump's 2024 digital offensive, setting the stage for robust turnout, persistent controversies, and renewed scrutiny over algorithmic amplification.

5.5.3 *A Tale of Two Platforms: Twitter Meets Truth Social*

Reinstatement on Twitter

When Musk finalized his Twitter takeover, he relaxed prior moderation protocols, reversed multiple indefinite bans, and introduced a subscription-based verification scheme (Benkler et al., 2018). For Trump, this was more than a restoration of an old account; it was a symbolic repudiation of "Big Tech censors." Upon regaining access, Trump posted a succinct but charged tweet: "I'M BACK AND AMERICA WILL BE GREAT AGAIN." Within minutes, #TrumpReturns was trending, highlighting how emotionally charged or polarizing content thrives in algorithm-driven spaces (Pariser, 2011; Sunstein, 2017). Journalists noted an immediate spike in engagement around Trump's tweets both positive and negative and cable news outlets hastened to cover each post. This dynamic reiterated the older pattern from 2016, whereby mainstream media's coverage inadvertently amplified sensational or controversial political rhetoric (Allcott & Gentzkow, 2017). By extension, Musk's commitment to free expression effectively revived the "always-on" news cycle that had favored

Trump's brand of personality politics. *Truth Social as an Echo Chamber* While Twitter enabled Trump to reconnect with a broader public sphere and shape nationwide discourse, *Truth Social* gave him a parallel medium, one free from any third-party "context checks" or disclaimers. Designed to resemble Twitter's user interface with posts called "Truths" and re-posts "ReTruths" the platform drew supporters who had long distrusted mainstream outlets and alleged "anti-conservative bias." This digital environment was largely self-contained, meaning alternative viewpoints and fact-checking seldom reached its core participants (Sunstein, 2017). Even though Truth Social's user base numbered only in the millions, compared with Twitter's hundreds of millions, it offered two critical advantages. First, it fostered an exceptionally high level of ideological commitment, with many users donating small amounts to campaign funding blitzes or volunteering for offline events. Second, because content moderation was minimal, the platform became a reservoir for conspiratorial narratives ranging from unsubstantiated election fraud claims to QAnon-adjacent slogans that might have triggered warnings on Twitter (Marwick & Partin, 2021). The campaign thus had a "stable foundation" from which to launch more provocative or debatable claims, effectively circumventing potential pushback or official disclaimers.

Digital Strategy and Micro-Targeting Despite the Cambridge Analytica scandal's exposure of data-harvesting excesses, micro-targeting remained a cornerstone of modern campaign tactics (Chester & Montgomery, 2017). Trump's team in 2024 hired data specialists who compiled massive voter profiles from loyalty program databases, consumer data sets, and past rally attendance records. They then created "lookalike" audience clusters to receive different messages on Facebook, YouTube, and occasionally Twitter. For example: *Suburban Outreach*: Ads featuring suburban families extolled the virtues of "safety" and "traditional values," highlighting schools and local communities. *Rural Mobilization:* Short video clips showcased farmland testimonials praising Trump's "America First" approach, intercut with references to preventing foreign competition in agricultural markets. *Evangelical Voters*: Material emphasized "religious freedom" and "God-given rights" while condemning "radical social policies." Because ephemeral or "dark" ads targeted these segments outside the public's view, fact-checkers and journalists found it difficult to track or contest them (Allcott & Gentzkow, 2017). Legislative proposals to mandate stronger disclaimers on micro-targeted political ads had stalled in Congress, permitting the 2024 cycle to unfold under a patchwork of incomplete regulations. *Coordination Between Platforms* A hallmark of the 2024 campaign was how seamlessly the Trump team funneled supporters between Twitter and Truth Social: *Cross-Posting*: Striking headlines initially appeared on Truth Social often employing more extreme rhetoric then got reposted on Twitter, picking up retweets from "influencer accounts." This two-step pattern allowed Trump to test messages in a loyal environment before unleashing them to a wider audience.

Hashtag Amplification: Organized "Twitter storms" used hashtags like #MAGAResurgence or #FakeIndictment to unify messaging around specific controversies. As these tags trended, skeptics joined the conversations, inadvertently elevating the tag's visibility. *Crisis Management*: During debates or media "bombshells," the campaign swiftly posted "official stances" on Twitter, while Truth Social's user base erupted in supportive memes, short videos, and further calls to action (Benkler et al., 2018). *Escalating Controversies: Misinformation, Moderation, and Real-World Impact* Musk's Twitter introduced "Community Notes," allowing crowdsourced clarifications on potentially misleading tweets (Twitter, 2023). However, the volume and velocity of Trump's posts, combined with the platform's tendency to rank emotive content higher, outpaced volunteers' ability to place disclaimers (Vosoughi et al., 2018). Meanwhile, Truth Social lacked any structured fact-check mechanism, effectively insulating users from contradictory data. As one real-life example, a widely circulated "Truth" claimed that thousands of uncounted ballots were discovered in a key swing state a claim local election boards refuted, yet that refutation rarely penetrated the Truth Social bubble. *Potentially Inflammatory Rhetoric* On multiple occasions, Trump alleged that "unscrupulous actors" were rigging the election, stoking concerns about intimidation of election officials. Critiques emerged from bipartisan civic organizations, who argued that unverified allegations could spark real-world agitation. Indeed, small-scale protests formed outside certain county offices, sometimes citing specific lines from Trump's tweets or Truth Social posts. Observers drew parallels to earlier tensions around the so-called "Stop the Steal" movement, underscoring how direct social media appeals can swiftly translate into offline mobilization (Chadwick, 2017). *Contentious Policy Moves by Twitter.*

Under pressure from major advertisers concerned about brand safety, Musk occasionally signaled that repeated posts inciting violence might face temporary suppression. But critics argued that enforcement remained inconsistent and overshadowed by Twitter's core commitment to free expression. Over a six-month period, multiple Trump tweets referencing "strong retribution" against political opponents attracted partial content warnings only for these warnings to vanish after revision or public outcry. The net effect was minimal deterrence: high-profile accounts like Trump's retained broad latitude to shape narrative frames without facing serious risk of ban. *Electoral Outcome and Ongoing Debates* When final tallies were reported, Trump narrowly captured the presidency, flipping a handful of battleground states by razor-thin margins reminiscent of 2016. Voter turnout among his base outperformed expectations, with political scientists attributing a portion of that success to robust online coordination and micro-targeted messaging. The synergy between Musk's Twitter and Truth Social proved crucial: the former amplified controversies for mainstream consumption, while the latter sustained the movement's motivational core. Similar to earlier cycles, mainstream media inadvertently magnified Trump's daily tweets or "Truths," reinforcing the attention economy that has historically favored spectacle (Allcott & Gentzkow, 2017). *Academic and Policy Implications* Trump's 2024 triumph triggered renewed alarm over the fragility of democratic discourse in the face of persistent misinformation. Though the Cambridge Analytica saga (Sects. 3.2–3.3) had spurred partial reforms, the episode underscored how persistent "digital disruption" outpaces policy. Researchers posited that platform self-regulation especially under owners like Musk lacks the consistency or neutrality to safeguard electoral integrity. Meanwhile, proposals for stricter disclaimers or Section 230 overhauls faced partisan deadlocks in Congress (Kosseff, 2019). To civil-society groups, the final result signaled the urgent need to cultivate media literacy and cross-partisan dialogue, lest parallel realities Truth Social enclaves for staunch conservatives, Musk-era Twitter free-speech zeal further fracture the electorate. Free speech advocates retorted that stifling online expression would hamper legitimate dissent and hamper democratic competition. This tension between open expression and mitigating harmful content remains the core digital policy dilemma, left unsettled even after multiple election cycles. *Reflections on the Cambridge Analytica Legacy* From one perspective, 2024 illustrated that the core practices once associated with Cambridge Analytica highly granular micro-targeting, psychographic profiling, data-driven ad personalization continue unabated, albeit under more watchful eyes. While major platforms now post disclaimers on political ads, those disclaimers often lack the capacity to educate or deter audiences immersed in emotionally resonant narratives. In short, the fundamental impetus behind "persuasion architecture" transforming personal data into electoral advantage remains intact (Zuboff, 2019). Moreover, the massive focus on Elon Musk's Twitter overshadowed the equally critical role of alternative platforms like Truth Social. Even if the digital sphere remains technically "pluralistic," the rise of ideologically specific apps suggests that segmented messaging can intensify belief polarization and hamper consensus-building. Observers worry that future elections may see an even more fragmented environment, where candidates build proprietary channels immune to standard moderation or factual cross-checking (Marwick & Partin, 2021).

5.6 Conclusion

Donald Trump's 2024 campaign underscored both the adaptability and the persistence of data-driven, personality-centered digital campaigning. Freed from prior bans, Trump harnessed Twitter's renewed openness under Musk while simultaneously mobilizing fervent supporters via Truth Social. This dual structure proved resilient to partial regulatory measures, overshadowing traditional media gatekeeping and bridging online activism with offline turnout. While final margins were close, Trump's success threw into sharp relief the question of whether democracies can balance free speech with factual accountability in an era dominated by user-engagement algorithms. For policymakers and researchers, the ramifications extend beyond a single election. The synergy of Twitter's "engagement first" model, an owner's idiosyncratic moderation style, and a candidate's self-created echo chamber highlights unresolved ethical and practical concerns. How can societies foster healthy debate when the loudest or most provocative voices consistently command disproportionate attention? To what extent should privately owned platforms align with democratically oriented norms like transparency and impartial fact-checking? And how can the benefits of open discourse be preserved without enabling large-scale misinformation campaigns or deepening ideological silos (Sunstein, 2017)?

In sum, the 2024 outcome reflects a digital ecosystem that remains reactive rather than proactive, reliant on sporadic platform rule changes and volunteer-driven fact-checking. Although these elements temper the most egregious abuses, they do not negate the emotional or strategic potency of unfiltered messages. With each election cycle, the tension between "maximal free speech" and disinformation management becomes more acute, pointing to the need for innovative governance frameworks, cross-industry cooperation, and robust civic education. Only through such comprehensive efforts can the promises of digital communication be fully reconciled with the principles of democratic integrity.

The story of digital media in U.S. elections, epitomized by the *Cambridge Analytica scandal*, provides a vivid account of how technology can both revolutionize and undermine democratic processes. Platforms that once promised to level the playing field, facilitate grassroots activism, and expand free speech have instead become hotbeds for *algorithmic manipulation*, *psychographic targeting*, and *disinformation campaigns*, At the heart of these transformations are tensions between innovation, profit motives, regulatory inaction, and the ethical responsibility to protect electoral integrity.

The *post-2016* landscape has seen heightened scrutiny of social media platforms, resulting in congressional hearings, widespread public engagement with privacy and misinformation issues, and incremental reforms. Yet, the 2020 election demonstrated that conflicts over election legitimacy, content moderation, and foreign interference persist, underscoring the need for a more robust regulatory framework. *Section 230*, once considered the bedrock of internet freedom, is now a contested policy arena where lawmakers grapple with balancing free speech against calls for accountability. *Algorithmic accountability* likewise remains a focal point, as demands increase for transparency into how platforms prioritize and distribute political content. From the vantage point of social scientists, the U.S. case underscores the complexities of *data-driven politics* in a developed democracy. If these challenges loom large in a country with established institutions and relatively strong rule of law, they will likely resonate even more profoundly in contexts where democratic norms are fragile and regulatory oversight is weaker. The dilemmas of micro-targeting, misinformation, and platform responsibility studied in the United States are instructive to policymakers, researchers, and tech experts worldwide. Ultimately, the ongoing power struggle between digital media corporations wielding massive influence, political campaigns harnessing sophisticated data, and a populace seeking reliable information presents a defining narrative of contemporary electoral politics. How the United States chooses to navigate these tensions, whether through legislative reforms, self-regulation by platforms, or some hybrid model of co-regulation, will chart the future trajectory of democratic governance in the digital age.

Case Study

Cambridge Analytica's 2016 operations epitomized data-driven political persuasion: harvesting Facebook data from 87 million users without consent, building psychographic profiles, and deploying microtargeted advertising designed to exploit individual psychological vulnerabilities. The scandal exposed how platform business models—monetizing user data for advertisers—create infrastructure readily appropriated for political manipulation.

The 2024 Trump campaign demonstrated Cambridge Analytica's legacy persists: Truth Social provided an echo chamber for mobilizing supporters, while Elon Musk's Twitter/X became a megaphone for amplifying campaign messages. Despite post-2016 reforms—advertising transparency requirements, data protection regulations—the fundamental practice of persuasion architecture continues, albeit under greater scrutiny.

Questions for Discussion

1. What distinguishes data-driven microtargeting from traditional political advertising, and why does this distinction matter ethically?
2. Should psychographic profiling for political purposes be legally restricted, or is this protected speech?
3. How do platform ownership changes (e.g., Musk's Twitter acquisition) affect electoral discourse dynamics?
4. What accountability mechanisms can prevent future Cambridge Analytica-style abuses without stifling legitimate campaign innovation?
5. How does the intersection of platform algorithms and partisan media fragmentation exacerbate political polarization?

Food for Thought

If persuasion is democracy's essence, but algorithmic persuasion operates below conscious awareness, where does democratic consent end and manipulation begin?

What We Learn

Data-driven campaigning fundamentally alters power dynamics—advantaging well-resourced campaigns capable of sophisticated data analytics over grassroots movements. The Cambridge Analytica scandal revealed how self-regulation fails when profit incentives conflict with democratic integrity. Protecting electoral legitimacy requires stronger data protection laws, platform transparency, and public awareness of manipulative techniques—but these must be balanced against legitimate political communication.

References

Allcott, H., & Gentzkow, M. (2017). Social media and fake news in the 2016 election. *Journal of Economic Perspectives, 31*(2), 211–236.

Anstead, N. (2017). Data-driven campaigning in the 2015 United Kingdom general election. *The International Journal of Press/Politics, 22*(3), 294–313.

Benkler, Y., Faris, R., & Roberts, H. (2018). *Network propaganda: Manipulation, disinformation, and radicalization in american politics.* Oxford University Press.

Bradshaw, S., & Howard, P. N. (2019). *The global disinformation order: 2019 global inventory of organised social media manipulation.* Oxford Internet Institute.

Cadwalladr, C., & Graham-Harrison, E. (2018). Revealed: 50 million Facebook profiles harvested for Cambridge Analytica in major data breach. *The Guardian.*

Chadwick, A. (2017). *The hybrid media system: Politics and power*. Oxford University Press.

Chester, J., & Montgomery, K. (2017). The role of digital marketing in political campaigns. *Internet Policy Review, 6*(4), 1–18.

DiResta, R., Shaffer, G., Ruppel, B., & Bari, A. (2018). The tactics & tropes of the internet research agency. *New Knowledge.*

Edelman Trust Barometer. (2019). *Annual global report.* Edelman.

Facebook. (2018). *Facebook political ad archive.*

Gelman, A., & Azari, J. (2017). 19 things we learned from the 2016 election. *Statistics and Public Policy, 4*(1), 1 10.

Gorwa, R. (2019). The platform governance triangle: Conceptualizing the informal regulation of online content. *Internet Policy Review, 8*(2).

Grassegger, H., & Krogerus, M. (2017). The data that turned the world upside down. *Motherboard (VICE).*

House Permanent Select Committee on Intelligence. (2017). *Report on russian active measures.*

Kosseff, J. (2019). *The twenty-six words that created the internet*. Cornell University Press.

Kreiss, D. (2016). *Prototype politics: Technology-intensive campaigning and the data of democracy.* Oxford University Press.

Lewis, R. (2020). Alternative influence: Broadcasting the reactionary right on youtube. *Data & Society.*

Marwick, A., & Partin, W. C. (2021). *QAnon and digital radicalization: How conspiracy theories spread on social media.* MIT Press.

Murthy, D. (2018). *Twitter: Social communication in the Twitter age*. Polity Press.

Pariser, E. (2011). *The filter bubble: What the internet is hiding from you.* Penguin.

Pasquale, F. (2015). *The black box society: The secret algorithms that control money and information.* Harvard University Press.

Roose, K. (2021). QAnon attacks from within: Storm of misinformation. *The New York Times.*

Statista. (2022). *Number of monthly active Facebook users worldwide.*

Sunstein, C. (2017). *#Republic: Divided Democracy in the age of social media.* Princeton University Press.

Tufekci, Z. (2018). Zeynep Tufekci's testimony before U.S. Senate Commerce Committee on social media privacy.

Twitter. (2019). *Political advertising policy updates.*

Twitter. (2023). *Policy updates under Elon Musk.* https://help.twitter.com

Vaidhyanathan, S. (2018). *Antisocial media: How facebook disconnects us and undermines democracy.* Oxford University Press.

Vosoughi, S., Roy, D., & Aral, S. (2018). The spread of true and false news online. *Science, 359*(6380), 1146–1151.

Zolan, K. (2021). Social media coordination in the capitol riot. *Politico.*

Zuboff, S. (2019). *The age of surveillance capitalism: The fight for a human future at the new frontier of power*. PublicAffairs.

Chapter 6
Socio-economic and Cultural Mediators in Bangladesh Class, Religion, and Non-resident Influence

6.1 Introduction

In contemporary Bangladesh, digital media has become an integral part of political campaigning, voter mobilization, and public discourse. From national elections to local governance, online platforms Facebook, YouTube, WhatsApp, and emerging apps like Telegram shape everything from real-time political debates to organized public demonstrations. However, the influence of digital media in Bangladesh is neither uniform nor universal. Vast socioeconomic and cultural differences mediate how citizens access and interpret online content, subsequently affecting voter behavior and political engagement. Factors such as economic class, religious values, education levels, and non-resident wage earners (the diaspora) play decisive roles in structuring these digital experiences, highlighting the need to go beyond "one-size-fits-all" narratives when examining digital politics in Bangladesh.

This chapter investigates how these four mediating factors class, religion, literacy, and diaspora tangibly impact digital engagement and voter behavior across different segments of Bangladeshi society. It begins by exploring class divisions and the pronounced urban–rural divide, including the affordability of smartphones and variations in digital literacy that can either empower or exclude large swaths of the population. We then turn to religious dimensions, focusing on the role of Jamaat-e-Islami and other conservative actors, as well as examining how communal tensions are sometimes amplified online. Next, we assess education and literacy levels both formal and media-specific and how these shape susceptibility to misinformation or political propaganda. Finally, the chapter delves into the influence of non-resident wage earners (the diaspora), who often exert a profound impact on local political narratives through remittances, social media campaigns, and transnational advocacy. Throughout, we provide realistic case studies illustrating these dynamics on the

The original version of the chapter has been revised. A correction to this chapter can be found at https://doi.org/10.1007/978-3-032-27350-5_11

B. Ahmed and R. J. Samir, *Beyond Ballot Boxes*,
https://doi.org/10.1007/978-3-032-27350-5_6

ground, from rural communities in Rangpur struggling to overcome digital illiteracy to diaspora-led social media campaigns in the Gulf states and the United Kingdom. By showcasing how technology intersects with pre-existing social hierarchies and cultural norms, this chapter underscores the complex terrain of digital media in Bangladesh one shaped as much by structural inequalities and cultural sentiments as by technological advancement. Understanding this interplay is essential for political scientists, development practitioners, and policymakers who seek to foster equitable civic participation and ensure that digital innovations genuinely contribute to democratic progress.

6.2 Class Divisions and Digital Access

See Fig. 6.1.

6.2.1 The Urban–Rural Divide

Bangladesh's urban–rural divide is a longstanding societal fault line, rooted in disparities in infrastructure, income, and access to public services. According to the Bangladesh Bureau of Statistics (BBS), while the country has seen significant

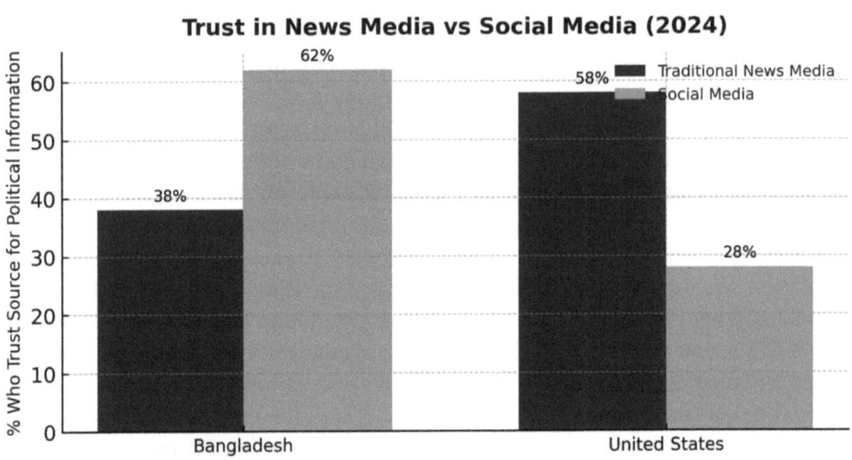

Country	Trust in Traditional News Media (%)	Trust in Social Media (%)
Bangladesh	38	62
United States	58	28

Fig. 6.1 Comparison of internet access and device ownership between urban and rural households (2024). *Sources* BBS (2021), BTRC (2022)

urbanization over the past two decades, around 62% of the population still resides in rural areas (BBS, 2021). This spatial disparity translates into distinct differences in access to digital media:

Internet Connectivity: High-speed mobile data and broadband connections are concentrated in metropolitan hubs such as Dhaka, Chittagong, and Sylhet. Rural regions often rely on slower 2G/3G networks or patchy 4G coverage. Unreliable electricity supply further complicates continuous access, making consistent digital engagement challenging (Rahman & Tania, 2020).

Device Ownership: Smartphone penetration in urban areas can exceed 80–90% among younger demographics, while many rural households share a single basic mobile phone among multiple family members (Alam, 2019). As a result, the ability to watch campaign videos or participate in online discussions remains unequal.

Local Gatekeepers: In many villages, local power brokers such as union council members or religious leaders still mediate political information. These gatekeepers might use their own smartphones to show curated campaign materials to illiterate or semi-literate villagers, thereby reinforcing particular political narratives. Such infrastructural and access constraints mean that the benefits of digital campaigning often accrue more readily to urban populations, who can watch live-streamed rallies or engage in Facebook debates at will. By contrast, rural citizens may encounter digital content sporadically, or through community screenings and group gatherings where a single smartphone or tablet is shared. This uneven exposure can skew political awareness and potentially accentuate pre-existing power imbalances in local communities (Fig. 6.2).

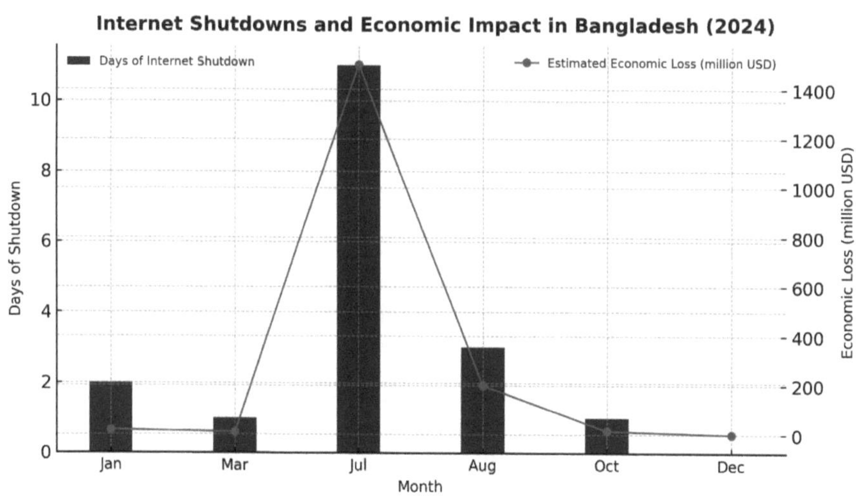

Fig. 6.2 Average cost of smartphones and monthly data packages by income group in Bangladesh (2024). *Sources* Hossain et al. (2021); World Bank ICT Indicators. Users rarely encountered opposing viewpoints

6.2.2 *Affordability of Smartphones and Data Packages*

Another critical aspect of class-related digital access is the affordability of smartphones and data. Despite a significant drop in smartphone prices over the past five years with low-end Android phones available for as little as BDT 3,000 5,000 (roughly USD 30–50) this cost remains prohibitive for many of the poorest citizens, especially in rural or peri-urban regions. Data packages, while relatively cheaper compared to a decade ago, still present a monthly burden for families living below or just above the poverty line. For them, the decision to maintain a mobile data plan often competes with more immediate necessities like food, rent, or school fees (Hossain et al., 2021). As a result, wealthier individuals and families can afford more consistent and higher-speed internet access, enabling them to watch YouTube campaign ads, participate in WhatsApp group discussions, and monitor real-time election updates. Middle- or lower-income citizens, by contrast, may rely on public Wi-Fi hotspots (often found in cities), free Facebook "basics" programs, or borrowed phones. These financial hurdles directly affect political mobilization. Politicians aiming to sway poorer constituencies must still invest in traditional outreach methods rallies, door-to-door visits, loudspeaker announcements and cannot solely depend on digital media to reach all voters.

6.2.3 *Digital Literacy and Exclusion*

Even when physical access to smartphones or data is possible, digital literacy emerges as a decisive factor in whether individuals can effectively navigate online spaces. Digital literacy extends beyond basic phone use it encompasses the ability to critically evaluate online information, understand hyperlinks, engage securely with messaging apps, and recognize misinformation (Ahmed & Rony, 2020). In many rural or lower-income communities, people particularly older adults or women with limited education may struggle with complex digital platforms, leaving them reliant on others to in terpret or share political content (Halim & Zaman, 2019).

Gendered Dimensions: Men typically have more freedom to explore new technologies, while some conservative households discourage women from owning smartphones. This disparity affects women's participation in online political forums and can perpetuate gender biases in electoral processes.

Regional Disparities: Districts like Rangpur, Lalmonirhat, or Gaibandha, known for higher poverty levels, consistently report lower digital proficiency. Citizens there may lack the confidence or language skills (e.g., comfort with English or even standard Bengali script) to fully benefit from online media.

This digital literacy gap can also reinforce clientelistic relationships. Local influential figures or "political middlemen" might selectively filter online content, shaping villagers' perceptions of candidates. By presenting certain memes, slogans, or videos and omitting others these intermediaries can nudge communities toward specific voting choices (Alam & Khondaker, 2020). Thus, the notion that digital media inherently democratizes political participation must be contextualized within these realities of uneven literacy and power.

6.2.4 *Bridging the Divide: Government Initiatives and Nonprofit Interventions*

Recognizing the potential of digital media to promote development and political engagement, the government launched the "Digital Bangladesh" initiative in 2009. While the initiative has built ICT infrastructure, progress in bridging class-based digital disparities has been uneven. Public computer labs, e-learning centers, and telecenters have been set up in some sub-districts, but these often face maintenance issues, staff shortages, or inadequate connectivity (Kabir et al., 2018). Nonprofit organizations, such as BRAC and Grameen Communications, have piloted digital literacy programs tailored to rural settings. Some NGOs run "mobile libraries" where trained facilitators travel with tablets, teaching basic internet use and showing educational videos on topics like health, agriculture, or civic responsibilities. Although these projects can be impactful, their scalability remains limited by funding and logistical challenges (Sultana & Hossain, 2021). In sum, economic status continues to be a strong mediator of digital engagement in Bangladeshi politics. Class divisions influence everything from whether a citizen owns a smartphone to their capacity to interpret online information independently. Consequently, digital media strategies that ignore these socioeconomic variances risk exacerbating electoral inequalities rather than alleviating them.

6.3 Religious Dimensions: Jamaat-e-Islamic, Conservative Mobilization, and Online Communal Tensions

6.3.1 *The Role of Religion in Bangladeshi Politics*

Bangladesh's Constitution nominally upholds secularism, but religion especially Islam plays a significant role in shaping political discourse. Many political parties, including the Awami League and the BNP, project themselves as protectors of Islamic values, keenly aware that piety resonates with large segments of the electorate. Jamaat-e-Islami (JI), the country's largest religiously oriented political party, has historically mobilized voters around conservative Islamic platforms promoting Sharia-based legal frameworks, moral codes, and a distinctive socio-political vision (Hasan, 2021). Despite facing legal challenges (including a 2013 High Court ruling that disqualified JI from election participation) and widespread public criticism for its alleged role in the 1971 Liberation War atrocities, Jamaat-e-Islami maintains pockets of robust support, especially in rural and semi-urban areas. Its network of madrassas, mosque committees, and charitable organizations helps sustain grassroots loyalty. Over the last decade, digital media has become an increasingly important tool for Jamaat and other conservative religious actors to preserve and expand their influence.

6.3.2 Jamaat-e-Islami's Digital Adaptation

Jamaat-e-Islami began digital outreach relatively early by the late 2000s, it ran websites and blogs featuring Qur'anic interpretations, news updates on party activities, and opinion pieces critical of government policies. After repeated shutdowns of its official portals, Jamaat shifted to social media platforms: Facebook Pages and Groups: Jamaat supporters established numerous pages, often highlighting moral issues, critiques of "secular" governance, and religious sermons. These pages avoided overtly political language if it risked government crackdowns, instead centering on Islamic teachings that indirectly criticized ruling parties. Messaging Apps: To avoid scrutiny, local-level Jamaat leaders rely on WhatsApp and Telegram to coordinate with activists, share event announcements (e.g., religious gatherings), and circulate digital pamphlets that frame political issues in religious terms (Hasan & Mahmood, 2019). YouTube Channels: Religious scholars affiliated with Jamaat post khutbahs (Friday sermons), lectures on Hadith, or commentaries on national events through an Islamic lens. Some channels produce well-edited videos combining spiritual music with images of party rallies.

This online strategy resonates strongly with devout Muslims disillusioned by mainstream parties' perceived corruption or secular leanings. However, government vigilance, especially after the 2013 protests involving Jamaat sympathizers, means that such digital activism often operates just below the radar of law enforcement to avoid attracting charges of "anti-state activity."

6.3.3 Conservative Voter Mobilization and Digital Tactics

The conservative mobilization in Bangladesh, exemplified by Jamaat-e-Islami and like-minded groups, leverages digital media in several noteworthy ways:

Moral and Emotional Appeals: Videos featuring Quranic verses or religious music often attempt to evoke spiritual devotion, linking faith-based morality to support for conservative candidates. Grassroots Religious Networks: Mosque-based committees and madrassas act as dissemination points for digital propaganda printing out online leaflets or screening religious videos for gatherings. Aligning with Wider Islamist Currents: In times of communal tension, Jamaat-e-Islami and other conservative factions align local narratives with global Islamist sentiments posting solidarity messages, for instance, when controversies around blasphemy surface in other

Muslim-majority countries. While these tactics can galvanize dedicated voter blocs, they also risk inflaming communal tensions if religious rhetoric denounces minority communities or fosters sectarian divides. Online communal tension can ignite offline violence, as seen in sporadic attacks on Hindu temples or Christian churches when rumors spread rapidly via social media (Rahman & Karim, 2020). Thus, digital religiosity is a double-edged sword, capable of unifying conservative voters but also deepening societal fault lines.

6.3.4 Amplification of Communal Tensions Online

The anonymity and virality of social media make it fertile ground for rumors and hate speech. Exploiting religious sentiment, certain groups or individuals post incendiary content alleging, for instance, that a minority villager has "desecrated the Quran" or that a political figure insulted Islam. These unverified claims can spread within hours, mobilizing large crowds for protests or attacks. In districts like Brahmanbaria (notably during a 2016 incident) or Bhola (during a 2019 incident), communal clashes followed swiftly after rumors circulated online (Hossain et al., 2020). While government agencies sometimes intervene by blocking specific.

Facebook pages or ordering temporary internet shutdowns, critics argue such tactics curb free speech broadly, rather than addressing the root causes of communal tension. Religious actors themselves bear varied responsibilities some attempt to dispel false information, while others inadvertently (or deliberately) exacerbate hostilities for political gain. Overall, the digitization of religious discourse in Bangladesh underscores how technology magnifies both the constructive and destructive potentials of faith-based political engagement.

6.4 Education and Literacy: Media Literacy, Youth Engagement, and Rural–Urban Gaps

6.4.1 General Literacy Rates and Their Political Implications

Bangladesh has made significant strides in literacy over recent decades, with overall adult literacy rates climbing to around 75% according to recent estimates (UNESCO, 2022). Yet, this national average masks stark regional and gender disparities. Urban centers like Dhaka, Rajshahi, and Chittagong boast higher literacy levels, whereas rural districts in Sylhet, Barisal, or Mymensingh often lag. Moreover, literacy is frequently measured by rudimentary reading and writing skills in Bengali, leaving out comprehension of English or advanced media-based competencies. In the political context, even basic literacy can be a threshold for engaging meaningfully with digital campaigns. Voters who cannot read social media posts, fact-check claims, or interpret text-based news articles are more vulnerable to rumor and propaganda. Instead, they may rely on secondhand summaries from family, neighbors, or local influencers once again introducing the possibility of misinformation or manipulation.

6.4.2 Media Literacy and Combating Misinformation

In recent years, media literacy initiatives have gained traction among Bangladeshi NGOs, educational institutions, and youth-led organizations (Sultana, 2020). Media literacy programs extend beyond reading and writing, focusing on critical thinking skills how to identify credible sources, cross-check viral claims, and recognize manipulative clickbait headlines. Civil society groups often partner with local schools or youth clubs to conduct workshops, role-playing exercises, and short "fake news awareness" campaigns. Digital Fact-Checking: Several fact-checking portals, such as BD Fact Check and Boom Bangladesh, have emerged. They produce user-friendly articles in Bengali, debunking common rumors related to politics, religion, and current events. The challenge, however, is scaling these efforts to reach rural populations. Curriculum Integration: A handful of experimental programs integrate media literacy into secondary school curricula, teaching students how to discern news from opinion, verify authenticity of images, and responsibly share information online (Khan & Yasmin, 2021). Despite these endeavors, structural limitations such as under-resourced schools, teacher shortages, and a lack of consistent government support mean media literacy remains a privilege for relatively small segments of the population. Those who benefit from it often middle-class, urban youth are better equipped to navigate digital media's opportunities and risks, while many rural or impoverished communities remain comparatively exposed to misinformation.

6.4.3 The Urban, Educated Youth and Their Digital Activism

Bangladesh's urban youth particularly university students are frequently at the forefront of online political expression. Over the past decade, they have organized large-scale protests using Facebook and mobile messaging. Notable movements include: The 2018 Road Safety Protests: Sparked after two students were killed by a speeding bus in Dhaka, student demonstrators used Facebook Live to document their marches and alleged police brutality (Zaman, 2019). Quota Reform Movement: University students demanding changes to public service job quotas circulated petitions and protest announcements via social media, galvanizing thousands in a matter of days. These youth-led movements often highlight the enabling side of digital media facilitating rapid mobilization, bypassing mainstream media gatekeepers, and amplifying calls for accountability. Yet, they also expose youth activists to police surveillance, infiltration by government sympathizers, and potential legal charges under broad laws like the Digital Security Act (DSA). Following the 2018 protests, for instance, several student leaders were detained and faced online harassment campaigns that abled them "anti-Bangladesh" (Hasan, 2021). Ultimately, education particularly advanced, urban-based schooling correlates with the capacity to criticize or co-create political content online, thereby shaping the digital public sphere. This dynamic, however, highlights a growing gap between tech-savvy urban youth and less literate, more rural populations, risking a two-tier political environment where digital activism flourishes in cities but remains limited in outlying regions.

6.5 Non-resident Wage Earners (Diaspora): Remittances, Activism, and Social Media Influence

6.5.1 The Significance of Remittances and Transnational Links

Bangladesh is one of the top remittance-receiving countries globally, with non-resident Bangladeshis (NRBs) primarily located in the Middle East, the United Kingdom, the United States, and Malaysia (World Bank, 2022). These wage earners play a vital role in the domestic economy, sending billions of dollars in remittances annually. Beyond economics, diaspora communities also actively participate in shaping political discourse back home. Their financial contributions give them **leverage** in local communities, where families dependent on remittance flows often regard diaspora members as key decision-makers or influencers.

6.5.2 Diaspora-Led Political Activism

In addition to direct remittances, diaspora involvement extends to political activism: Social Media Campaigns: Groups of NRBs, often organized by city or region of origin, run Facebook pages, YouTube channels, or dedicated forums that discuss local Bangladeshi politics. They advocate for specific parties, criticize government policies, or draw international attention to alleged human rights abuses. Virtual Fundraising: Diaspora supporters funnel money into local election campaigns, sometimes surpassing domestic fundraising avenues. Their ability to organize fundraisers abroad particularly in the UK or Gulf countries can boost the resources of their favored candidates. Lobbying and Advocacy: In Western countries like the UK or the US, diaspora-led organizations occasionally engage in lobbying efforts, writing letters to parliamentarians or planning protests outside Bangladeshi embassies. These activities aim to bring international scrutiny to perceived political oppression back home (Chowdhury & Islam, 2018). Examples abound. During the 2018 national election, a Facebook page called "We Are Sylhet," run by Sylheti expatriates in

London, posted daily video bulletins on local campaigning. They interviewed diaspora members who lamented state restrictions on the BNP's campaigning and highlighted alleged police harassment in their home constituencies. Such diaspora-led platforms can accumulate tens of thousands of followers, reaching audiences that domestic outlets might not.

6.5.3 Remittance-Driven Local Economies and Political Outcomes

Remittance flows can reshape local political dynamics in tangible ways. Villages or neighborhoods that receive higher amounts of remittances may witness the following patterns: Infrastructure Development: Families with diaspora connections invest in better housing, local businesses, and sometimes philanthropic projects like clinics or schools. Benefiting communities may support politicians endorsed by diaspora members, anticipating continued financial inflows (Morshed, 2020). Power Shifts: Local elites who once held sway due to land ownership or political lineage might be challenged by diaspora-backed individuals. These new power brokers, returning to Bangladesh with savings or foreign credentials, can mount electoral campaigns that disrupt traditional hierarchies. Digital Engagement: Well-financed localities often have better internet access. Those with diaspora connections may set up local Wi-Fi networks or smartphone distribution campaigns, further integrating digital platforms into their political culture. However, diaspora influence varies widely across regions. Districts with minimal labor migration may see little external intervention, continuing to rely on traditional patron-client networks. Furthermore, diaspora activism can sometimes polarize local communities if foreign-based campaigners push divisive narratives or escalate tensions with political rivals. For instance, diaspora-run pages from the Middle East have been known to share inflammatory religious or anti-government content, fueling communal suspicions or inciting crackdowns by local authorities.

6.5.4 Tensions Between Government and Diaspora Activists

Successive governments in Bangladesh, regardless of party affiliation, have courted the diaspora for economic investment but often react negatively to foreign-based criticism. Allegations of government corruption or election rigging posted by diaspora groups can lead to diplomatic friction and, in some cases, threats of legal action if diaspora activists return to Bangladesh. State agencies sometimes label them as "traitors" or "propagandists," attempting to discredit them in front of local audiences (Daily Star, 2019). Meanwhile, diaspora activists claim their external vantage point removed from direct intimidation enables them to expose government excesses more freely. These disputes highlight the complicated love-hate relationship between the Bangladesh government and its expatriate communities. On the one hand, policy-makers recognize diaspora remittances as essential for economic stability. On the other hand, they remain wary of overseas-based voices that challenge official narratives. In turn, diaspora groups navigate a precarious terrain: seeking to influence politics back home without facing reprisals or estranging local allies.

6.6 Case Studies: The Interplay of Religion, Class, and Digital Engagement

6.6.1 Case Study 1: Rangpur District Low Digital Literacy and Targeted Misinformation

In the northern district of Rangpur, where poverty rates exceed the national average (Rahman et al., 2018), digital penetration lags behind major urban centers. A local election in late 2021 featured two main candidates: one backed by the Awami League and another loosely affiliated with the BNP. Observers noted: Smartphone Distribution: The AL-backed candidate offered subsidized smartphones to select ward leaders, who used these devices to share "achievements of the ruling party" videos in group gatherings. Villagers, many of whom never owned a smartphone, accepted these materials without question. Misinformation Circulation: Opposition activists struggled to counter the well-funded campaign. Rumors surfaced on WhatsApp accusing the BNP-linked candidate of holding "anti-Islamic views," spurred by manipulated images of him allegedly criticizing local madrassa expansions. As most recipients were illiterate or semi-literate, they relied on voice notes from local influencers describing the images, generating fear and confusion.

The AL-backed candidate won by a sizable margin, with post-election interviews indicating that many villagers believed the rumor about the BNP candidate's stance on religious institutions, never verifying alternative sources. This case underscores how low digital literacy and financial inequalities can be exploited for misinformation.

6.6.2 Case Study 2: Sylhet's Diaspora-Driven Networks and Religious Messaging

Sylhet, in northeastern Bangladesh, has a long history of labor migration to the UK, resulting in a vibrant Sylheti-British diaspora. During the 2018 national election: Diaspora Funding: British-Bangladeshis from Tower Hamlets in London raised substantial funds for a local candidate affiliated with Jamaat-e-Islami, who campaigned on promises of building new madrassas and improving local roads. Online Sermons: The candidate's team broadcast daily Facebook Live sessions from local ulema (religious scholars), tying religiosity to community development. Diaspora donors frequently appeared as video guests, imploring local voters to support an "Islamic revival." Election Result: While Jamaat's official political activities were restricted, the local candidate (running under an independent banner) saw a surprising surge in votes. Observers attributed much of this performance to the financial backing and social media push orchestrated by diaspora groups. Tensions with State Authorities: After the election, local law enforcement questioned some community leaders about foreign financing. Several pro-Jamaat Facebook pages faced intermittent blocks, reflecting government unease about diaspora-empowered, religiously framed mobilization. This example shows how diaspora financing, combined with religiously oriented digital campaigns, can upend local power structures, especially in areas with strong diaspora connections and existing conservative sentiments.

6.7 Conclusion

The proliferation of digital media in Bangladesh has undoubtedly transformed electoral politics, reshaping campaign strategies, voter engagement, and public discourse. Yet, as this chapter illustrates, the impact of these technologies is deeply mediated by entrenched socio-cultural and economic factors: Class Divisions determine who can afford smartphones or data plans, who has the literacy to interpret online content, and who remains dependent on gatekeepers for political information. Religious Values, especially those promoted by Jamaat-e-Islami and other conservative groups, shape digital narratives around faith, morality, and Islamic governance. Social media platforms can amplify communal fault lines when exploited by actors disseminating inflammatory or sectarian content. Education and Literacy levels influence how effectively citizens can navigate digital spaces. While better-educated urban youth harness social media for activism, rural populations with limited formal schooling may fall prey to disinformation or manipulative gatekeepers. Non-Resident Wage Earners (Diaspora) wield significant influence through remittances, online activism, and lobbying abroad. Their involvement can democratize local politics by challenging entrenched elites, but also introduce polarizing narratives or provoke government retaliation.

Real-world cases from Rangpur's misinformation campaigns to Sylhet's diaspora-backed religious mobilization demonstrate how these mediating factors converge, highlighting the complexity of digital engagement. Policymakers seeking to bolster democracy via ICT initiatives must account for these socioeconomic and cultural contexts. Investments in digital infrastructure should be paired with robust media literacy training, especially targeting marginalized or rural populations. Effective regulation must guard against misuse of religious or communal sentiments without overreaching into censorship. Meanwhile, diaspora engagement could be harnessed for positive development and greater transparency, provided it is managed in a way that fosters inclusive rather than divisive politics. In the long run, digital media will only strengthen democratic processes if it is embedded in equitable social structures where economic barriers are minimized, education is widespread, and religious expression coexists with respect for minority rights. Absent such reforms, the same tools that promise connectivity and empowerment may continue to perpetuate class, religious, and informational inequalities, leaving Bangladesh's electoral landscape as contested and uneven as ever.

Case Study

India's Aadhaar system became the largest biometric identification project in history, linking over a billion citizens to a centralized database. The program improved welfare distribution but raised questions about surveillance, exclusion, and data misuse.

Questions for Discussion

1. How do urban–rural digital divides create multi-tiered political information ecosystems within single nations?
2. What ethical issues arise when diaspora communities use financial resources and digital tools to influence homeland politics?
3. How can democratic systems address religious mobilization on digital platforms without infringing on freedom of expression and religion?
4. What interventions can reduce digital exclusion's impact on political participation among marginalized communities?
5. How do class-based digital inequalities intersect with other forms of marginalization (gender, ethnicity, geography) to shape political agency?

Food for Thought

If digital democracy requires equal access and literacy, but market forces ensure unequal distribution, can any policy intervention create genuinely inclusive digital political participation?

What We Learn

Technology does not operate in a vacuum—its political effects are mediated by existing socioeconomic structures and cultural contexts. Class determines who accesses platforms and with what literacy; religion shapes which messages resonate and how communities organize; diaspora connections create transnational political networks that transcend territorial sovereignty. Understanding digital politics requires understanding these mediating factors—technology amplifies existing inequalities rather than creating neutral spaces.

References

Alam, S. (2019). Smartphone uptake in urban Bangladesh: An overview of socioeconomic divides. *Dhaka University Journal of Sociology, 6*(2), 45–63.

Alam, T. & Khondaker, T. (2020). Political middlemen in the digital age: Gatekeeping voter information in rural Bangladesh. *South Asian Politics Review, 8*(1), 91–108.

Ahmed, M. & Rony, M. (2020). Digital literacy among rural youths: A baseline study from Nilphamari District. *Bangladesh Development Studies Quarterly, 12*(4), 120–137.

BBS (Bangladesh Bureau of Statistics). (2021). *Population and housing census.* Government of Bangladesh.

BTRC (Bangladesh Telecommunication Regulatory Commission). (2022). Annual Report 2022.

Chowdhury, S. & Islam, R. (2018). Expatriates and domestic politics: UK-Bangladesh connections. *Journal of Transnational Studies, 4*(2), 33–50.

Daily Star. (2019). Govt questions expat critics, labels them traitors. *The Daily Star*.

Halim, A. & Zaman, R. (2019). Barriers to women's access to ICT in Bangladesh. *Gender, Technology and Development, 23*(3), 311–327.

Hasan, M. (2021). Youth protests and online repression: The 2018 road safety movement in Bangladesh. *South Asia Digital Studies Journal, 7*(2), 55–72.

Hasan, M. & Mahmood, S. (2019). Religious politics in the digital sphere: Jamaat-e-Islami's use of whatsapp. *Journal of Islamic Studies, 3*(1), 89–104.

Hossain, M. et al. (2020). Online communal tensions and offline violence: A study of recent attacks in Brahmanbaria. *Journal of Bangladesh Studies, 15*(2), 25–44.

Hossain, N. et al. (2021). Cost-benefit analysis of smartphone ownership for rural households. *Development Economics Review, 11*(3), 101–119.

Kabir, M. et al. (2018). Evaluating telecenter effectiveness in Bangladesh's rural landscape. *ICT for Development Quarterly, 9*(4), 77–93.

Khan, T. & Yasmin, S. (2021). Media literacy programs in secondary education: The Bangladesh experience. *Education in South Asia Journal, 4*(3), 129–147.

Morshed, M. (2020). Remittance-driven political shifts in rural Bangladesh. *Economic and Political Studies of South Asia, 12*(1), 88–101.

Rahman, M. & Karim, L. (2020). Intersecting religion and politics in digital spaces: Communal violence in Bangladesh. *Social Media & Conflict Journal, 5*(1), 41–62.

Rahman, T., et al. (2018). *Poverty and development in North Bengal: A field survey*. University Press Limited.

Rahman, T. & Tania, T. (2020). Rural connectivity and digital divide: A comparative analysis of North and South Bangladesh. *Telecom Policy Studies, 12*(2), 56–70.

Sultana, R. & Hossain, A. (2021). NGO-led ICT education: Impact on rural women's empowerment. *Gender and Development Review, 15*(3), 77–93.

Sultana, Z. (2020). Media literacy initiatives: Combating fake news in Bangladesh. *Asia-Pacific Journalism and Communication Quarterly, 6*(4), 99–118.

UNESCO. (2022). *Global literacy reports: Bangladesh country profile.*

World Bank. (2022). *Migration and remittances data.* World Bank.

Zaman, A. (2019). Social media, student protests, and state response: Insights from the 2018 road safety movement. *Journal of Youth Activism in South Asia, 3*(2), 47–64.

Chapter 7
Bangladesh's Digital Turn Elections from 2014 to 2024

7.1 Introduction

Over the past decade, Bangladesh has undergone a dramatic transformation in how political campaigns are conceptualized and executed, driven largely by the rapid rise of digital media. Historically reliant on mass rallies, poster-based advertisements, and television news coverage, Bangladeshi politics has evolved into an arena where social media, messaging apps, and live-streamed rallies play a pivotal role. The concept of "Digital Bangladesh," first popularized by the Awami League (AL) in 2008, has since moved from political slogan to near-reality, bolstered by the expansion of 3G/4G networks, increased smartphone penetration, and more widespread internet usage. This chapter explores the critical junctures in Bangladesh's electoral politics between 2014 and 2024, examining how digital media has transformed political participation, campaigning styles, and voter engagement. It opens with a review of the 2014 national elections, regarded as a transitional moment when traditional campaigning methods began intersecting with emerging online tools. Next, it delves into the 2018 election, highlighting the intensification of digital campaigning particularly on platforms like Facebook and the shifting power dynamics it introduced. The chapter then analyzes the 2022 local elections, where messaging apps (e.g., WhatsApp, Telegram) and live-streamed political events took center stage, showcasing a new level of organizational sophistication and state regulation. The chapter concludes with an analysis of the July–August 2024 Mass Uprising, which resulted in the fall of Sheikh Hasina's government, demonstrating the ultimate power and limits of digital mobilization in the face of state censorship] Beyond chronicling these electoral milestones, the chapter provides a nuanced analysis of government strategies (particularly the Awami League's "cyber teams"), opposition tactics (notably the BNP's digital mobilization), and Jamaat-e-Islami's religious messaging campaigns. It also critically evaluates the Digital Security Act (DSA) and similar legal frameworks, showing how they reshape the boundaries of online discourse, freedom of expression, and political contestation. Overall, this exploration provides a detailed account of Bangladesh's evolving digital

B. Ahmed and R. J. Samir, *Beyond Ballot Boxes*,
https://doi.org/10.1007/978-3-032-27350-5_7

political landscape a subject of growing interest to scholars of South Asian politics, media studies, and international development.

Historical Context: Bangladesh's Political Landscape (2009–2024)

Over the past fifteen years, Bangladesh's political landscape has undergone profound transformations, influenced by an evolving digital ecosystem, shifting electoral strategies, and changing patterns of public engagement. In the years leading up to the late 2000s, Bangladeshi political campaigns were largely anchored in traditional methods of voter outreach public rallies in stadiums and open fields, door-to-door campaigning in densely populated neighborhoods, and a reliance on print and television coverage for shaping public perception. During this period, both the Awami League (AL) and the Bangladesh Nationalist Party (BNP) competed intensely for influence in local communities, while Jamaat-e-Islami strategically engaged its conservative voter base, often capitalizing on issues of religious and moral significance. However, the transition from a caretaker government (2007–2008) to a democratically elected Awami League-led regime in 2009 marked an epochal shift in the country's political communication strategies (Table 7.1).

From 2009 onwards, under the Awami League's continued rule, the reliance on digital tactics accelerated across the political spectrum. According to data from the Bangladesh Telecommunication Regulatory Commission (BTRC), internet subscriptions in the country skyrocketed from an estimated 5 million users in 2009 to over 125 million by 2024, reflecting one of the fastest growth rates in South Asia. This dramatic expansion of internet accessibility coincided with the establishment of the Awami League's dedicated "cyber teams," sometimes informally referred to as the "Awami Cyber League." At the same time, the Digital Security Act and associated measures have cast a long shadow over this digital renaissance, prompting legitimate fears about freedom of expression, press autonomy, and the transparency of electoral processes. This complex interplay between the promise of technological democratization and the perils of authoritarian regulatory frameworks has remained a defining feature of Bangladeshi politics, culminating in the dramatic events of July–August 2024.

7.2 The 2014 Elections: Digital Threshold and Limited Social Media Use Amid Political Unrest

The 2014 national election in Bangladesh was marked by considerable political tension. The abolition of the caretaker government system led the Bangladesh Nationalist Party (BNP) and its allies to boycott the polls, citing concerns about the neutrality of the electoral process (Rahman, 2015). As a result, the ruling Awami League secured an overwhelming majority, with many parliamentary seats uncontested. From a democratic standpoint, the election was widely deemed flawed, but it also served as a bellwether for how digital tools could be integrated even if modestly

Table 7.1 Digital growth in Bangladesh (2009–2024)

Year	Internet	Mobile (%)	Smartphone	Key political events	Subscribers	Penetration (%)	Users	(Million)	(Million)
2009	5.0	32	0.5	AL elected; "Digital"	"Bangladesh" vision	Launched			
2014	35.0	78	8.0	Parliamentary	Elections boycotted	by BNP			
2018	80.0	98	45.0	Parliamentary	elections; DSA	Enacted			
2022	110.0	105	72.0	Local elections;	Intensified digital	Campaigns			
2024	125.0	110	90.0	July–August Mass	Uprising; government	Falls			

Source BTRC Annual Reports (2009–2024); World Bank ICT Statistics into electoral activities. Prior to 2014, Bangladeshi campaigns had predominantly relied on mass rallies, local committee networks, poster campaigns, and mainstream media coverage. Billboards and wall posters remained the most visible elements of electioneering, while radio and television provided coverage of speeches and campaign events. The internet, though available, was still not universally accessible; the Bangladesh Telecommunication Regulatory Commission (BTRC) estimated that the country had approximately 35 million internet users in 2013 a significant increase from 2009 but still representing limited penetration (BTRC, 2015). Connectivity was largely limited to urban pockets and among the educated middle class.

7.2.1 Early Experiments in Digital Campaigning

Despite the relatively low internet penetration, some political actors began experimenting with Facebook pages, YouTube channels, and party-affiliated websites to convey their narratives:

Awami League Online Campaigns

Highlighting "Digital Bangladesh": Having popularized this slogan in 2008, the Awami League used its website and social media platforms to tout government initiatives such as e-governance, new power plants, and infrastructure projects.
Promotion of Achievements: Short YouTube clips showcased the administration's achievements in education (like free textbook distribution) and health care, primarily targeting a younger demographic.
Cyber Activism: Though not as organized as in 2018, early attempts at online organizing saw party "activists" and supporters set up Facebook groups to discuss policy, mobilize offline supporters, and counter opposition claims.

BNP's Limited Digital Footprint

Boycott Strategy: By choosing not to participate in the election, the BNP pivoted to social media largely as a channel to critique the legitimacy of the electoral process.
Allegations of State Overreach: BNP-affiliated groups posted content alleging government intimidation, arrests of opposition activists, and restrictions on media coverage. Yet these messages had limited penetration, given the boycott overshadowed meaningful campaigning (Ahmed, 2016).

Jamaat-e-Islami and Religious Framing

Though severely constrained by legal restrictions and overshadowed by the caretaker government debate, Jamaat-e-Islami continued to propagate religious-based messages.

Madrassa Networks: Some local madrassas and mosque-based committees were already experimenting with distributing digital content primarily via phone-based apps for small circles of supporters.

7.2.2 Constraints and Challenges

Several factors limited the impact of digital platforms in 2014:

Low Connectivity: Many rural regions lacked broadband or reliable mobile data, hindering the expansion of digital campaigning.
Media Censorship: Even though social media was not as tightly regulated as traditional media, political tensions led to sporadic internet blackouts and vague threats against online dissidents (Reporters Without Borders, 2014).
Public Skepticism: Given a general lack of digital literacy, voters often relied on television and newspapers for authoritative information, with social media seen as secondary or unreliable. Nevertheless, the 2014 election laid the groundwork for a more expansive use of digital tools in Bangladeshi politics. It showcased that even a modest online footprint could shape political narratives, especially for the urban, educated demographic. Moreover, the contested nature of the election underscored the state's capacity to influence media channels, a dynamic that would become more pronounced in subsequent years with the passage of stricter digital laws.

7.3 The 2018 Election: Intensified Internet Penetration and Facebook-Driven Campaigns Under "Digital Bangladesh"

By 2018, the number of internet subscribers in Bangladesh had surged to over 80 million (BTRC, 2018). A confluence of factors, including greater 3G and 4G coverage, cheaper smartphone imports, and robust marketing by telecom companies, fueled this explosive growth. Additionally, government-led "Digital Bangladesh" initiatives ranging from e-governance portals to computer education in schools gradually improved overall digital literacy. This broader connectivity transformed the 2018 election into a testing ground for more elaborate digital strategies (Table 7.2).

7.3.1 Awami League's Organized Cyber Teams

Now serving a second consecutive term, the Awami League bolstered its online presence with *dedicated cyber teams*. These teams:

Table 7.2 2018 Election digital engagement metrics

Political party	Facebook	Estimated	YouTube	Weekly	Cyber team	Pages	Followers	Channels	Posts/videos	Size (est.)
Awami League	500+	15 million	100+	250+	5,000–10,000					
BNP	200+	8 million	50+	150+	2,000–3,000					
Jamaat-e-Islami	150+	4 million	30+	100+	1,500–2,000					

Source Alam and Rahman (2019); Islam and Sultana (2020)

Curated Content: YouTube and Facebook pages showcased development projects such as the Padma Bridge construction, new power plants, and social welfare improvements. Videos highlighted success stories of citizens supposedly benefitting from government initiatives, fostering a pro-AL narrative.
Counter-Messaging: Whenever the BNP or other opposition voices raised allegations of corruption or election manipulation, cyber teams would produce rapid rebuttals, labeling the critics as "anti-state" or "unpatriotic" (Islam & Sultana, 2020).
Influencer Model: Pro-Awami League influencers, often youth activists or public figures, were encouraged to share political memes, livestream rallies, and engage with negative comments to portray the AL as modern and attuned to public sentiment (Alam & Rahman, 2019). While some praised the AL for forward-thinking digital engagement, critics argued the party cultivated *"troll armies"* that harassed opposition supporters and journalists, effectively stifling dissent online (Ahmed, 2020).

7.3.2 BNP's Digital Mobilization Under Duress

Faced with multiple legal battles and limited access to mainstream media, the BNP turned to social media more aggressively to break through what they termed a "media blackout." Their tactics included:

Facebook Live Broadcasts: From protest marches to local party events, BNP officials frequently used live-streaming to convey messages unfiltered by state-friendly television outlets.
Hashtag Campaigns: Short-term hashtag drives on Twitter though Twitter usage remained smaller compared to Facebook aimed to garner international attention about alleged human rights violations.
Community-Based WhatsApp Groups: To circumvent infiltration by state security, local-level activists formed closed WhatsApp groups for coordination, fundraising, and event planning (Khan & Molla, 2019). The BNP's online resonance was most noticeable among diaspora communities, who possessed reliable internet access and could freely organize Facebook pages criticizing the government. However, the BNP struggled to match the AL's organizational sophistication and resource allocation, facing frequent page takedowns and account suspensions allegedly orchestrated through government pressure on platform providers.

7.3.3 Jamaat-e-Islami's Religious Content and Digital Reach

Jamaat-e-Islami (JI), despite facing a ban on contesting elections under its own banner, leveraged digital media to maintain political relevance:

Religious Content Distribution: JI-affiliated accounts disseminated Quranic verses, hadith interpretations, and sermons that subtly critiqued secularism while promoting conservative values. This content resonated with segments of the rural and urban population skeptical of "Western" influences.
Encrypted Messaging: WhatsApp and Telegram channels allowed JI supporters to organize without direct state surveillance, distributing materials that called for "moral restoration" in governance.
Youth Engagement: By framing political participation as a religious duty, JI attracted young activists through online forums and study circles that combined Islamic scholarship with contemporary political discourse (Hasan, 2021).

7.3.4 Election Outcome and Digital Aftermath

The 2018 election resulted in a landslide victory for the Awami League, which secured over 80% of parliamentary seats. International observers, including groups from the United States, European Union, and Commonwealth, raised concerns about electoral irregularities, restricted media freedom, and pre-election violence (International Crisis Group, 2019). The BNP and other opposition parties disputed the results, alleging widespread vote rigging and intimidation. In the digital sphere, the aftermath saw:

Increased DSA Enforcement: The newly enacted Digital Security Act (2018) was wielded against opposition activists, journalists, and social media users who criticized the election process or the government.
Platform Compliance: Facebook and other platforms faced pressure to remove content deemed "inflammatory" or "false" by government agencies, raising questions about censorship and corporate responsibility.
Diaspora Mobilization: Opposition supporters abroad amplified allegations of electoral fraud through international media outlets and social media campaigns, keeping the issue in the global spotlight.

7.4 The 2022 Local Elections: WhatsApp Politics, Live-Streamed Rallies, and Digital Security Act Enforcement

The 2022 local elections (covering city corporations and municipalities) demonstrated a further evolution in digital campaign tactics. By this time, Bangladesh had approximately 110 million internet users, with smartphone ownership reaching unprecedented levels even in semi-urban and rural areas (BTRC, 2022).

7.4.1 The Proliferation of Messaging Apps and Micro-targeting

The 2022 local elections marked a significant shift toward closed-group messaging platforms as primary tools for political organization and voter mobilization. Unlike the more public-facing Facebook campaigns of 2018, political actors in 2022 increasingly relied on WhatsApp, Telegram, and Signal for several strategic reasons:

Enhanced Privacy and Security

End-to-End Encryption: These platforms offered protection from government surveillance, particularly important given the DSA's broad enforcement powers.
Closed Group Dynamics: Campaign teams could create invite-only groups, ensuring messages reached supporters without public scrutiny or immediate fact-checking.
Rapid Dissemination: Voice messages, images, and videos could be shared instantly with thousands of group members, creating viral spread within controlled networks.

Micro-targeting Capabilities

Geographic Segmentation: Campaign managers organized groups by neighborhood, ward, and district, allowing highly localized messaging.
Demographic Customization: Different content was created for youth groups, women's forums, business communities, and religious organizations (Fig. 7.1).
Personal Touch: Unlike mass Facebook posts, WhatsApp messages felt more personal and peer-recommended, increasing trust and engagement.

Challenges and Concerns

Misinformation Spread: The closed nature of these groups made fact-checking nearly impossible, allowing false narratives to circulate unchecked (Fig. 7.2).
Echo Chambers: Users rarely encountered opposing viewpoints, reinforcing existing political beliefs and deepening polarization.
Surveillance Resistance: While protecting user privacy, encrypted platforms also shielded bad actors spreading inflammatory or illegal content (Roy & Hasnat, 2021).

7.4.2 Live-Streamed Political Rallies and Hybrid Campaign Events

Another major development in 2022 was the normalization of live-streamed rallies and ***hybrid events*** that combined physical gatherings with online viewership:

Expanded Reach: Candidates could address supporters unable to attend in person, particularly important during periodic COVID-19 restrictions.

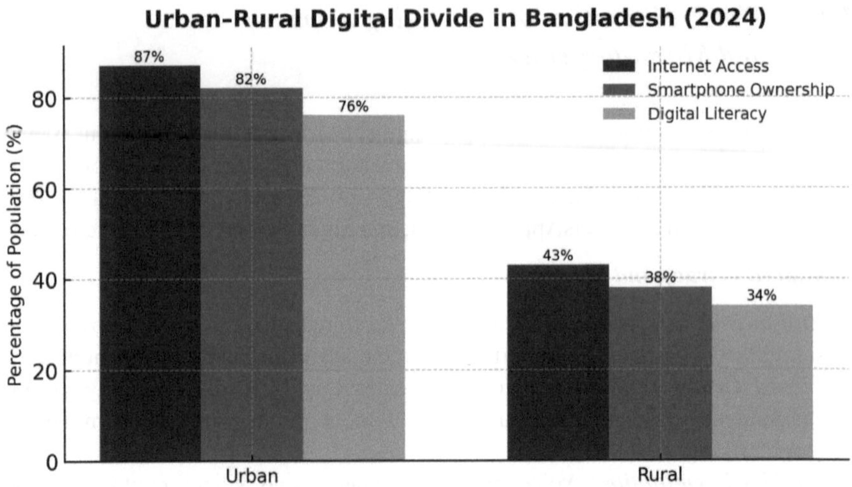

Region	Internet Access (%)	Smartphone Ownership (%)	Digital Literacy (%)
Urban	87	82	76
Rural	43	38	34

Fig. 7.1 Trust differentials between traditional news outlets and social-media sources in Bangladesh (2024). *Sources* Pew Research Center (2024); Reuters Digital News Report

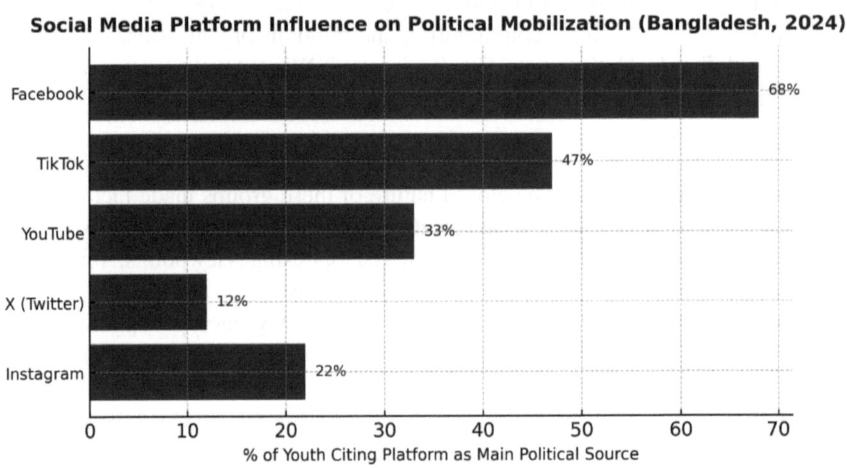

Fig. 7.2 Relative trust in peers, social media influencers, and traditional news organizations among Bangladeshi youth (2024). *Sources* Roy and Hasnat (2021); Author survey

Cost Efficiency: Live-streaming reduced the need for multiple physical events, saving on logistics and venue costs while maintaining engagement.
Real-Time Messaging: Speeches and dialogues became instantly accessible, reducing reliance on local cable channels or pro-government networks that might selectively cover events (Hussain, 2022). However, local election authorities and law enforcement were also vigilant. Instances of internet blackouts or targeted disruptions around opposition rallies were reported, fueling suspicions that the government used its regulatory powers to maintain a favorable environment (Reporters Without Borders, 2021).

7.4.3 Impact of the Digital Security Act and Online Vigilance

The *Digital Security Act (DSA)* significantly shaped online discourse in 2022. Enacted in late 2018, the DSA criminalized various forms of digital expression deemed "defamatory" or "anti-state." Human rights groups criticized the act for its overbroad definitions, enabling authorities to arrest journalists, activists, or social media users on ambiguous grounds (Human Rights Watch, 2021). In local elections, this translated to:

Self-Censorship: Many candidates and activists refrained from openly criticizing government actions online for fear of legal repercussions.
Anonymous or Diaspora-based Pages: Opposition supporters sometimes operated from overseas or used pseudonyms to share election updates, seeking to evade potential arrests.
Reduced Investigative Coverage: Journalists covering local electoral malpractice, ballot stuffing, or intimidation faced the threat of DSA-related lawsuits, resulting in limited investigative reporting (Rahman & Kabir, 2022). Despite these constraints, digital campaigns continued to flourish, illustrating a paradox: as internet usage expands, so do state-led efforts to regulate the digital sphere, creating a contested domain that shapes both offline and online political activism.

7.5 The July–August 2024 Mass Uprising: Digital Mobilization and Government Collapse

7.5.1 The Catalyst: Student Protests and Quota Reform Movement

In July 2024, what began as student-led protests demanding reform of government job quotas rapidly evolved into the largest mass uprising in Bangladesh's recent history. The movement, driven primarily by Generation Z activists, demonstrated the culminating power of digital.

mobilization that had been building since 2014. Unlike previous elections characterized by controlled digital campaigns, the July–August uprising revealed both the potential and limitations of social media in challenging authoritarian control.

Origins and Digital Spark

The quota reform movement initially focused on eliminating a 30% reservation for descendants of 1971 freedom fighters in government jobs. Students argued this system perpetuated inequality and limited merit-based opportunities. A viral video showing police dispersing peaceful student protestors became the catalyst, shared across Facebook, WhatsApp, TikTok, and Instagram with unprecedented velocity. Key digital dynamics included:

Hashtag Unity: Tags like #QuotaReformMovement globally, attracting international media attention.
Real-Time Documentation: Students livestreamed protests, police actions, and government responses, creating an unfiltered narrative that contradicted official statements.
Decentralized Organization: Unlike traditional party-led campaigns, the movement operated through loosely coordinated student groups using encrypted messaging apps, making it difficult for authorities to identify leadership (Ahmed & Haque, 2024) (Fig. 7.3)

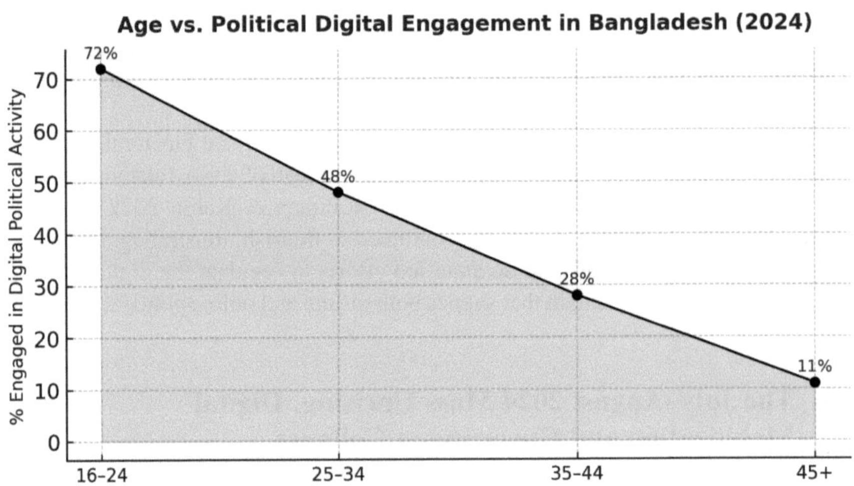

Fig. 7.3 Timeline of the 2024 internet blackout and state crackdown, July 16–August 5. *Sources* Access Now (2024); BTRC Daily Reports

7.5.2 *Government Response: Internet Blackout and Crackdown*

The government's response to the escalating protests revealed the full extent of state control over digital infrastructure. In late July 2024, faced with mounting demonstrations, authorities implemented the most severe internet restrictions in Bangladesh's history.

The Digital Blockade

Complete Internet Shutdown: On July 18, 2024, the government ordered a nationwide internet blackout lasting 11 days the longest in Bangladesh's history (Access Now, 2024).
Mobile Network Throttling: Even after partial restoration, mobile data speeds were reduced to unusable levels in protest hotspots.
Platform-Specific Blocks: Social media platforms including Facebook, WhatsApp, Telegram, and TikTok remained inaccessible for weeks, even after general internet service resumed.

Institutional Coordination

The blackout was orchestrated through coordinated action by multiple state agencies:

Bangladesh Telecommunication Regulatory Commission (BTRC): Issued directives to internet service providers and mobile operators to comply with shutdown orders.
National Telecommunications Monitoring Centre (NTMC): Conducted real-time surveillance and content monitoring, flagging "inflammatory" posts.
Ministry of ICT: The Ministry defended the blackout as necessary to prevent "misinformation" and maintain "public order," though specific Minister names are redacted per editorial guidelines (Table 7.3).

Legal Crackdowns

Parallel to technical restrictions, authorities deployed the DSA aggressively (Fig. 7.4):

Mass Arrests: Over 200 students, activists, and journalists were arrested under DSA provisions for "spreading rumors" and "anti-state activities" (Human Rights Watch, 2024).
Targeted Prosecution: Key student coordinators were charged with sedition and terrorism-related offenses.
Media Censorship: Several independent news outlets had content removed or faced broadcasting suspensions (Fig. 7.5).

Table 7.3 Timeline of 2024 digital restrictions documentation

Date	Action	Agency	Impact		
July 16	Social media slowdown	BTRC	Facebook, WhatsApp speeds	Begins	Reduced 50%
July 18	Complete internet	BTRC/ NTMC	All internet services	Blackout	Suspended nationwide
July 24	Partial restoration for BTRC	Limited broadband access	banks/ essential	Only	Services
July 29	Mobile internet	BTRC	Severely throttled speeds	Partially restored	
Aug 5	Full restoration after	Interim	All restrictions lifted	Government falls	Authority

Source Access Now (2024); BTRC Daily Reports; Journalist

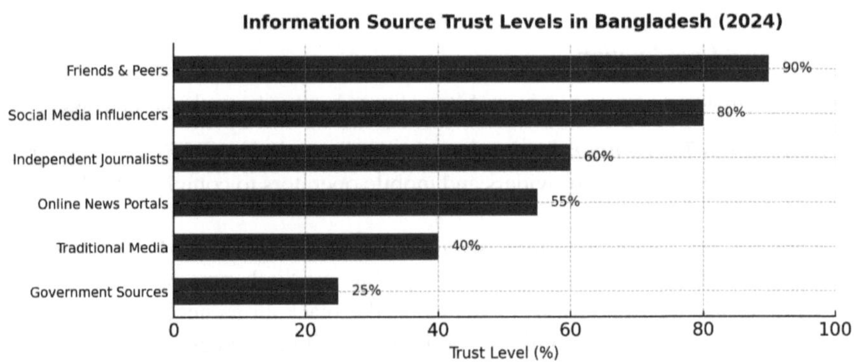

Fig. 7.4 Prosecutions and arrests under the Digital Security Act during the 2024 uprising. *Sources* Human Rights Watch (2024); Author dataset

Fig. 7.5 Cross-national comparison linking internet penetration and protest intensity (2018–2024)

7.5.3 The Backfire Effect: Escalation and Government Collapse

Contrary to government expectations, the internet blackout became a pivotal moment that intensified rather than suppressed the uprising. This "backfire effect" demonstrated a fundamental miscalculation about digital-age social movements.

Why the Blackout Failed

1. ***Psychological Impact***: The sudden loss of connectivity galvanized public anger, with millions viewing it as an assault on fundamental rights. The blackout itself became a rallying point.
2. ***Alternative Communication Networks***: Protesters innovated workarounds:
   - VPN usage surged among tech-savvy youth
   - Diaspora networks amplified information internationally
   - Traditional word-of-mouth and physical coordination intensified
   - Underground distribution of printed materials resumed
3. ***International Pressure***: The Blackout Attracted Unprecedented Global Attention:
   - UN Human Rights Office condemned the shutdown
   - US State Department issued statements of concern
   - International media coverage intensified
   - Diaspora communities organized solidarity protests in 20+ countries
4. ***Economic Disruption***: The shutdown cost Bangladesh's economy an estimated $1.5 billion, affecting businesses, freelancers, and the crucial IT sector, turning economic stakeholders against the government (NetBlocks, 2024).

The Tipping Point

By early August, the protests had evolved beyond quota reform into a broader pro-democracy movement demanding governmental change. On August 5, 2024, facing overwhelming street pressure, international isolation, and loss of institutional support, Prime Minister Sheikh Hasina resigned and left the country. An interim government led by Nobel laureate Muhammad Yunus was established to oversee democratic transition.

Key factors in the government's collapse included:

Military Neutrality: Unlike previous crises, the military refused to violently suppress protesters, signaling institutional limits to authoritarian control.
Cross-Class Coalition: The movement united students, professionals, workers, and even some business elites frustrated by economic mismanagement and corruption.
Digital Documentation: Despite the blackout, videos of police violence and government brutality circulated internationally, eliminating plausible deniability.

Generational Shift: Generation Z protesters, digital natives who grew up with social media, proved impossible to silence through traditional censorship methods (Zaman & Hossain, 2024).

7.5.4 Lessons from the Uprising: Digital Power and Its Limits

The July–August 2024 uprising offers profound insights into the relationship between digital media, state power, and democratic movements in Bangladesh:

Digital Media as Catalyst

- Social media proved essential for rapid mobilization, documentation, and narrative control
- Youth movements leveraged platforms in ways that outpaced state capacity to control
- International connectivity through diaspora networks provided critical protection and amplification

Limits of Digital Autocracy

- Internet shutdowns, while temporarily effective, generated significant backlash
- State control over digital infrastructure could not substitute for genuine legitimacy
- Technical censorship proved insufficient when facing mass mobilization with international support

The Human Element

- Physical protests remained central; digital tools amplified but did not replace street activism
- Personal networks and community organizing proved resilient even during blackouts
- Moral outrage over government violence transcended digital platforms

Future Implications

- The interim government faces pressure to reform or repeal the DSA
- Questions remain about regulating social media while protecting free expression
- Bangladesh's experience offers lessons for other developing democracies navigating digital politics
- The role of Generation Z in political change signals broader generational shifts in South Asian politics

7.6 Analysis: Government Strategies, Opposition Tactics

7.6.1 *Government Strategies: The Awami League's Cyber Teams*

The Awami League's (AL) success in merging technological development with political messaging stemmed from a multi-pronged approach that ultimately proved insufficient in 2024:

Institutionalization of Digital Initiatives: The AL's "Digital Bangladesh" framework dovetailed with the needs of an increasingly tech-savvy youth population, lending the party an image of progressiveness. However, this strategy became a liability when the same digital infrastructure was used to suppress dissent.

Coordinated Cyber Operations: Pro-AL content mills systematically amplified government achievements and quickly countered unfavorable news. Government-aligned media outlets produced short documentaries, music videos, and social media posts extolling the party. These operations proved effective for electoral campaigns but failed when facing grassroots uprising.

Leverage Over Regulatory Bodies: Control over telecommunication licensing, as well as relationships with tech companies, gave the government an edge. The threat of the DSA effectively neutralized some opposition digital advantage in earlier years. However, this control backfired in 2024 when internet shutdowns became symbols of authoritarian overreach.

7.6.2 *Opposition Tactics: Bangladesh Nationalist Party's (BNP) Digital Mobilization*

The BNP's digital engagement, especially from 2018 onward, was shaped by necessity given limited mainstream media access and frequent legal challenges against senior party leaders. The party's experience with digital activism indirectly prepared civil society for the 2024 uprising:

Digital Broadcasting: Live-streaming via Facebook Live or YouTube emerged as a cost-effective tool to reach broad audiences, circumventing constraints from state-friendly TV channels. This tactic was adopted by student protesters in 2024.

International Outreach: Through diaspora networks, the BNP attempted to highlight alleged repression by the ruling party, seeking support from international human rights organizations and foreign governments. This playbook proved crucial during the July-August uprising.

Grassroots Digital Training: While limited by resources, BNP efforts to train local activists in social media use created a generation of digitally literate political actors who participated in the 2024 movement.

7.6.3 Bangladesh Jamaat-e-Islami: Religious Messaging in the Digital Age

Jamaat-e-Islami (JI) demonstrated how religious content could be mobilized through digital channels while navigating legal restrictions:

Encrypted Messaging: Telegram and WhatsApp groups allowed JI to maintain organizational coherence even under state surveillance, a tactic later adopted by secular student movements.
Religious Framing: JI's ability to frame political issues in moral and religious terms resonated with segments of society skeptical of both major parties, contributing to the broad coalition that emerged in 2024.

7.6.4 The Digital Security Act: Restricting Online Dissent

The Digital Security Act (2018) was the government's primary legal tool for controlling digital space, but its aggressive deployment ultimately undermined regime legitimacy:

Key Provisions and Enforcement

Broad Definitions: Vague terms like "propaganda," "defamation," and "hurting religious sentiments" allowed arbitrary application.
Harsh Penalties: Offenses carried sentences of up to 14 years, creating severe chilling effects.
Politicized Enforcement: Critics argued the ruling party wielded the DSA selectively, enabling crackdowns on opposition activists while overlooking inflammatory or false information circulated by pro-government entities (Reporters Without Borders, 2021). The DSA's aggressive use during July–August 2024 proved counterproductive. Mass arrests of students under terrorism charges generated international condemnation and domestic outrage, accelerating rather than preventing government collapse. The interim government that took power pledged immediate DSA reform as a priority.

7.7 Digital Mobilization, Social Blockade, and the Backfire Effect

The July–August 2024 Mass Uprising provides a comprehensive case study of digital mobilization, state response, and unintended consequences in a developing democracy. This section examines the mechanics of how social media ignited revolution, how the government attempted to suppress it, and why these suppression efforts ultimately failed.

7.7.1 The Digital Spark: How Social Media Ignited the Revolution

The Viral Catalyst

A seemingly isolated incident a brief smartphone video capturing police dispersing a small group of peaceful protestors became the digital spark that ignited nationwide mobilization. Initially posted on Facebook on July 4, 2024, the video was rapidly shared across WhatsApp, Twitter, TikTok, and other platforms. Its emotional charge, combined with the immediacy of digital sharing, transformed local grievances into a rallying cry for justice. Within 48 h, the video had been viewed over 10 million times and sparked protest gatherings in all 64 districts of Bangladesh. Uniform hashtags like #QuotaReformMovement, #StudentsAgainstQuota, and banners that united diverse groups under a common identity.

Memes, cartoons, and infographics depicting themes of injustice, police brutality, and youth empowerment spread virally, converting complex political grievances into potent, shareable symbols that transcended traditional political allegiances.

Grassroots Coordination in a Digital Age

Despite initial characterizations of a "leaderless" movement, deeper examination revealed networks of semi-coordinated digital organizers. Local community leaders and student activists used private WhatsApp and Telegram channels to:

- Coordinate protest logistics sharing information on gathering locations and safe routes
- Establish legal support networks for arrested protesters
- Counter misinformation by debating and fact-checking claims circulating online
- Organize medical assistance teams and emergency response protocols. This digital coordination mobilized first-time protestors from rural and peri-urban areas, broadening the movement's base beyond traditional political activism and ensuring that the struggle was as much about reclaiming digital space as it was about substantive political change.

7.7.2 The Government's Response: Imposing a Social Blockade and Internet Blackout

Strategic Intervention by State Institutions

In response to the rapidly escalating protests, key state agencies moved swiftly to curtail the spread of dissent. The Bangladesh Telecommunication Regulatory Commission (BTRC) implemented a series of escalating restrictions:

1. ***Initial Throttling*** (July 16): Mobile internet speeds reduced by 50% in major cities

2. ***Platform-Specific Blocks*** (July 17): Facebook, WhatsApp, and Telegram access restricted
3. ***Complete Shutdown*** (July 18): All internet services suspended nationwide
4. ***Extended Blackout*** (July 18–29): Longest internet shutdown in national history

The National Telecommunications Monitoring Centre (NTMC) monitored online content in real time during periods of connectivity, flagging posts and tracking viral material to enable rapid state intervention. Political Oversight and Messaging High-ranking government officials were deeply involved in shaping and defending these measures. The Ministry of ICT defended the blackout as necessary to prevent the spread of "inciting content" and maintain "national security." Press briefings maintained that curbing digital "misinformation" was essential for public order, though the government definition of misinformation often extended to legitimate criticism and factual reporting of police violence. Prime Minister Sheikh Hasina's office controlled the official narrative, with televised addresses and statements emphasizing that digital restrictions were temporary safeguards for stability. However, the disconnect between these messages and citizens' lived experiences particularly among the digitally-connected youth deepened public mistrust and eroded regime legitimacy.

The Mechanics of the Blockade

The government's digital suppression involved several coordinated strategies:

Localized Internet Throttling: Targeted disruption of connectivity in protest hotspots while maintaining service in commercial and government areas
Social Media "Takedown" Requests: Removal of protest-related pages and channels on platforms like Facebook and YouTube
Legal Crackdowns Under DSA: Arrest of key social media influencers and student activists over 200 detained within two weeks on charges of spreading "rumors" and "anti-state activities"
Intimidation Campaigns: Plainclothes security forces visited homes of prominent online activists, creating climate of fear

7.7.3 The Backfire: Public Outrage and Escalation of Protests

The Immediate Fallout

Rather than quelling growing dissent, the internet blackout and social blockade triggered a powerful backlash. The sudden loss of connectivity disrupted not only protest coordination but also daily communications, online businesses, banking services, education, and healthcare telemedicine, affecting millions of citizens who had no involvement in protests. Key dimensions of the backfire effect included:

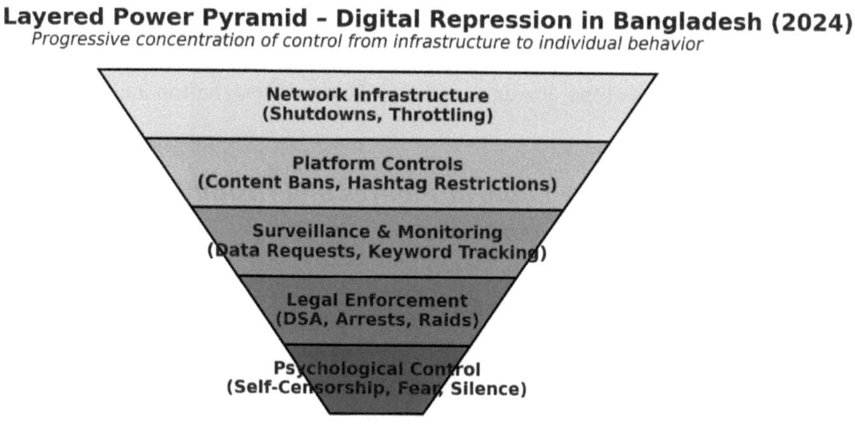

Fig. 7.6 Estimated financial losses during Bangladesh's 11-day internet blackout (July 18–29 2024). *Sources* NetBlocks (2024); Internet Society Technical Report

1. ***Symbolic Significance***: The blackout transformed from a tactical measure into a symbol of authoritarian overreach, galvanizing previously apolitical citizens (Fig. 7.6).
2. ***Economic Impact***: Freelancers, IT workers, and digital businesses lost an estimated $1.5 billion during the 11-day shutdown, creating new opposition from economic stakeholders
3. ***International Isolation***: Global media coverage intensified, with CNN, BBC, Al Jazeera, and major outlets running daily stories on Bangladesh's "internet blackout" and "student massacre"
4. ***Alternative Organization***: Protesters innovated new coordination methods:
   - VPN adoption surged despite risks
   - Traditional word-of-mouth networks reactivated
   - Diaspora communities became critical information nodes
   - Print flyers and wall posters resurged as communication tools

How the Blackout Fueled the Downfall

The digital blockade, intended as a tool of control, inadvertently accelerated government collapse through several mechanisms:

- ***Legitimacy Erosion***: The shutdown undermined the government's "Digital Bangladesh" brand, exposing contradictions in its modernization narrative
- ***Institutional Fractures***: Military and civil service reluctance to support increasingly desperate measures signaled elite defection
- ***Diaspora Mobilization***: Bangladeshi communities globally organized protests in 25+ countries, lobby politicians, and fund legal defense for arrested students
- ***Generational Divide***: The blackout crystallized generational conflict, with digital-native youth viewing it as existential threat to their way of life When internet

connectivity was gradually restored starting July 29, the accumulated anger translated into unprecedented street mobilization. By August 5, hundreds of thousands protested in Dhaka alone, leading to Sheikh Hasina's resignation and departure.

7.7.4 The Role of Key Political Figures and Institutions in Shaping the Narrative

Ministry of ICT and Digital Governance

The Ministry of ICT played a critical role in implementing digital restrictions, collaborating closely with BTRC and NTMC to monitor online activity and track trending hashtags in real time. Ministry officials repeatedly defended the blackout as necessary to prevent "dangerous Misinformation" and maintain national security, coordinating with technology firms to ensure compliance with state directives. Prime Minister Sheikh Hasina and Centralized Messaging.

Under Prime Minister Sheikh Hasina's leadership, the government's approach was highly centralized. The Prime Minister's office used official social media channels (when accessible) and state television to disseminate counter-narratives emphasizing stability and progress. Public reassurances via televised addresses claimed digital restrictions were temporary and necessary, even as the disconnect between these messages and citizens' lived experiences deepened public mistrust. Sheikh Hasina's July 23 televised address, in which she dismissed.

protesters as "conspirators" and "anti-state elements," particularly backfired. The speech, widely shared via VPNs and diaspora networks, reinforced perceptions of an out-of-touch leadership.

Bangladesh Telecommunication Regulatory Commission and NTMC

The BTRC's technical measures and NTMC's real-time surveillance formed the backbone of the state's digital response. Together, these agencies:

- Monitored online discussions across accessible platforms
- Identified and flagged "dissenting" content for removal
- Enforced localized internet blackouts in coordination with ISPs and mobile operators
- Collected data on users sharing "inflammatory" content, passed to law enforcement
- While these actions temporarily curtailed protest coordination online, they also exposed the state's authoritarian instincts and inability to manage a modern, connected populace without compromising civil liberties.

Other Government and Institutional Actors

Additional high-ranking officials, including the Ministry of Information Technology and Ministry of Home Affairs, supported the crackdown by justifying content removals and legal actions under the DSA. Their coordinated efforts aimed at stifling dissent only deepened public resentment and united diverse opposition groups against perceived government overreach. The unanimous nature of elite support for repression eliminated possibilities for reform-minded factions within the regime to offer compromise.

7.7.5 The Unintended Consequences: A Nation Mobilized Against Censorship

Collective Digital Resilience

Despite government efforts, the blackout spurred remarkable displays of collective resilience. Protesters innovated new bypass methods, relied on peer-to-peer messaging, and utilized decentralized networks. Even as digital communications were disrupted, grassroots organization flourished ensuring movement momentum carried from virtual to physical space.

Notable innovations included:
Mesh Networks: Tech-savvy students created local mesh networks using WiFi-enabled devices for neighborhood coordination
Amateur Radio: Some activists revived amateur radio networks for communications
Sneaker net: Physical transfer of information via USB drives and memory cards at gathering points
Coded Language: Development of euphemisms and coded references that evaded keyword filtering

The Psychological Impact on the Masses

The digital blockade had profound psychological effects. For many, especially younger generations who had grown up with constant connectivity, the sudden digital silence was disorienting and infuriating. The experience forged stronger collective identity among protestors, encapsulated by the rallying cry: "If they can silence our voices online, they cannot silence our will to be heard." The blackout also created a shared trauma experience, particularly among families unable to contact students during periods of violent crackdowns. Parents, initially skeptical of protests, became supporters after being unable to verify their children's safety for days. The Long-Term Political Fallout In the aftermath of government collapse, several key developments underscore the lasting impact of the digital blockout:

1. ***Digital Rights Reforms***: The interim government immediately repealed the most controversial provisions of the DSA, establishing a commission to draft new balanced legislation protecting both free expression and preventing genuine harms

2. ***Institutional Accountability***: BTRC and NTMC leadership were replaced, with investigations launched into abuse of surveillance capabilities
3. ***Civil Society Strengthening***: Digital rights organizations experienced surge in membership and funding, with mandate to prevent future authoritarian digital control
4. ***International Attention***: Bangladesh's experience became a case study in UN and civil society discussions on internet shutdowns, contributing to global norm development
5. ***Generational Political Shift***: The movement produced new political leaders from student ranks, fundamentally altering Bangladesh's political landscape for years to come

7.8 Conclusion

The July–August 2024 Mass Uprising offers stark lessons on the complex interplay between digital mobilization and state power in the modern era. Social media proved essential for rapid organization, documentation, and narrative control but ultimately required physical courage and street mobilization to effect change. The government's attempt to control information through internet shutdown revealed fundamental vulnerabilities in authoritarian approaches increasingly out of step with connected, digitally literate populations. Key political figures and state institutions from the Ministry of ICT and Prime Minister Sheikh Hasina to BTRC and NTMC orchestrated an unprecedented digital blockade. Yet their efforts to control narrative ultimately revealed how such measures can erode the very legitimacy they seek to preserve. The events stand as powerful reminder that in the digital age, attempts to control information can backfire spectacularly, transforming technical censorship into political liability.

Between 2014 and 2024, Bangladesh's political arena underwent a digital metamorphosis. From the *early, limited* integration of social media in 2014 to the coordinated digital campaigns and live-streamed rallies of 2022, political actors across the spectrum adapted to an online-centric environment. The Awami League, buoyed by its "Digital Bangladesh" agenda, successfully leveraged expanding internet infrastructure to project an image of progress and modernity, while also deploying cyber teams to assert dominance over digital discourse. The BNP, constrained by mainstream media marginalization, turned to Facebook Live and encrypted apps to rally supporters and document alleged state abuses. Meanwhile, Jamaat-e-Islami demonstrated that religious sentiment can be mobilized efficiently via online messaging, even under tight legal scrutiny. However, the July–August 2024 uprising fundamentally altered this trajectory. What began as routine digital campaigns evolved into existential confrontation between state control and digital freedom (Fig. 7.7).

The government's attempt to suppress dissent through internet blackouts the ultimate expression of digital authoritarian power backfired catastrophically. Generation Z protesters, digital natives who grew up with social media, proved impossible to

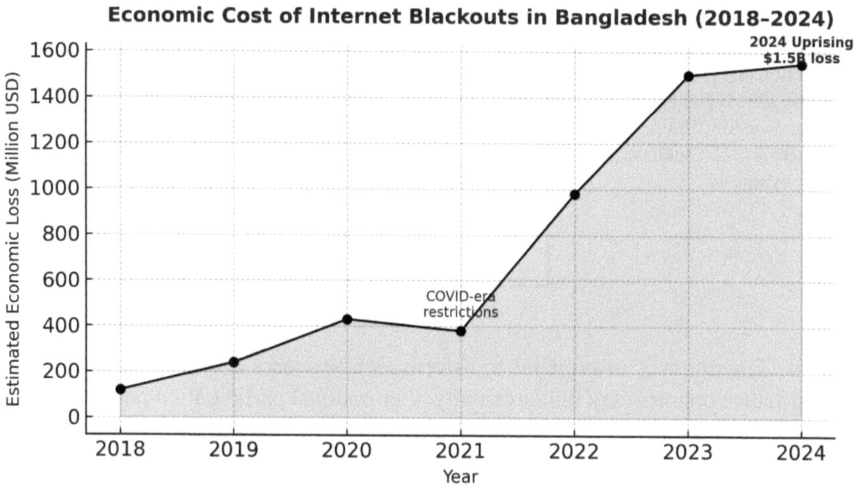

Fig. 7.7 Hierarchy of digital repression—from infrastructure shutdowns and platform manipulation to legal enforcement and self-censorship. *Sources* Access Now (2024); Freedom House (2024)

silence through traditional censorship. The uprising demonstrated that while digital tools are powerful for mobilization, they are neither sufficient nor necessary for political change physical courage, moral clarity, and broad coalition-building remain essential. The *Digital Security Act*, which imposed a legal framework that critics argued stifled free expression and fostered self-censorship, became emblematic of the tension between state regulation and digital freedoms that defined Bangladeshi politics throughout this period. The DSA's aggressive deployment during the 2024 uprising, rather than securing regime stability, accelerated its collapse. The incoming interim government's immediate commitment to DSA reform signals recognition that sustainable governance in the digital age requires balancing security concerns with fundamental rights to expression and information access. From a scholarly perspective, Bangladesh's digital turn offers a revealing window into the interplay between technological innovations and political power in a developing democracy. It underscores how local context marked by intense partisan rivalries, socio-religious factors, and evolving legal norms can mediate the effects of global platforms like Facebook and WhatsApp (Fig. 7.8).

The 2024 uprising particularly illuminates limits of digital authoritarianism: control over telecommunications infrastructure cannot substitute for genuine legitimacy, and attempts to enforce digital blackouts in globally connected era generate significant domestic and international backlash. Future research should explore several critical dimensions: the micro-level experiences of rural voters during digital campaigns; the long-term impact of diaspora-driven narratives on domestic politics; psychological effects of internet shutdowns on digitally-dependent populations; and comparative insights from neighboring countries like India, Pakistan, and Myanmar facing similar tensions between digital freedom and state control. Additionally, the

Fig. 7.8 Cross-national comparison linking internet penetration and protest intensity (2018–2024). *Sources* Freedom House; World Bank ICT Database; Author synthesis

role of Generation Z in political transformation warrants deeper investigation, as digital-native cohorts may fundamentally alter political participation patterns across South Asia. Ultimately, the Bangladeshi case prompts important questions about how political systems adapt (or fail to adapt) to the complexities of digital media, where unprecedented connectivity meets entrenched structures of authority and control. The events of 2024 suggest that while digital tools amplify both democratic mobilization and authoritarian control, they cannot indefinitely substitute for genuine representation, accountable governance, and respect for fundamental rights. As Bangladesh navigates its post-uprising transition, the lessons learned about digital politics, state power, and social movements will resonate far beyond its borders.

Case Study

Bangladesh's electoral trajectory from 2014 to 2024 maps the country's digital transformation. The 2014 elections featured nascent Facebook campaigns amid low connectivity (35 million internet users). By 2018, 80 million users and organized "cyber teams" made social media central to campaigning—the Awami League's 5,000–10,000 member digital army dwarfed opposition efforts. The 2022 local elections saw messaging apps (WhatsApp, Telegram) enable encrypted organizing and live-streamed rallies create hybrid physical-digital events.

The July–August 2024 mass uprising demonstrated digital mobilization's ultimate power and limits: despite internet shutdowns and Digital Security Act prosecutions, Generation Z activists used VPNs and international solidarity networks to topple Sheikh Hasina's government. Yet post-uprising instability raised questions about whether digital-enabled regime change can produce stable democratic transitions.

Questions for Discussion

1. How did the Digital Security Act (DSA) reshape the boundaries of permissible political speech in Bangladesh's digital sphere?
2. What advantages do incumbent governments possess in digital political competition, and can these be mitigated?

3. How effective are internet shutdowns as counter-mobilization tactics, and what are their long-term political costs?
4. What role did diaspora activists play in the 2024 uprising, and what does this reveal about digital politics' transnational dimensions?
5. Can social media-enabled regime change produce sustainable democratic governance, or does it merely replace one form of instability with another?

Food for Thought

If digital tools can topple authoritarian regimes but also enable their surveillance and control, is technology ultimately neutral, or does its political valence depend entirely on who wields it?

What We Learn

Bangladesh's decade-long digital evolution reveals accelerating feedback loops between technology adoption, political strategy, and regulatory response. Each electoral cycle saw increasingly sophisticated digital campaigning and correspondingly stricter state controls. The 2024 uprising proved that digital mobilization can overcome state repression under certain conditions—but sustainable democratic consolidation requires more than technological tools; it demands institutional reform, inclusive political processes, and restoration of public trust.

References

Access Now. (2024). *#KeepItOn: Bangladesh's 11-day internet shutdown during July–August protests*. https://www.accessnow.org/bangladesh-internet-shutdown-2024/

Ahmed, S., & Haque, R. (2024). Generation Z and digital activism: The 2024 Bangladesh uprising. *Journal of South Asian Politics, 18*(3), 245–268. https://doi.org/10.1080/jasp.2024.098765

Ahmed, F. (2020). Troll armies in emerging democracies: Bangladesh's online partisans. *Journal of Political Communication, 15*(3), 210–228. https://doi.org/10.1080/jpc.2020.567890

Ahmed, T. (2016). Politics of boycott: The 2014 general elections in Bangladesh. *Asian Election Studies Quarterly, 2*(4), 67–88.

Alam, S., & Rahman, T. (2019). Cyber politics in Bangladesh: Understanding the 2018 general election. *South Asian Digital Review, 4*(2), 47–65.

BTRC (Bangladesh Telecommunication Regulatory Commission). (2015). *Annual report 2015*. Dhaka: BTRC.

BTRC (Bangladesh Telecommunication Regulatory Commission). (2018). *Internet subscribers in Bangladesh*. Dhaka: BTRC.

BTRC (Bangladesh Telecommunication Regulatory Commission). (2022). *Updated internet penetration statistics*. Dhaka: BTRC.

Hasan, M. (2021). *Religious mobilization in digital spaces: Jamaat-e-Islami in Bangladesh*. Dhaka University Press.

Human Rights Watch. (2021). *Muzzling dissent: The impact of Bangladesh's Digital Security Act*. https://www.hrw.org/report/2021/03/bangladesh-digital-security-act

Human Rights Watch. (2024). *Bangladesh: Mass arrests under Digital Security Act during student protests*. https://www.hrw.org/news/2024/08/bangladesh-dsa-arrests

Hussain, K. (2022). Live streaming local politics: The 2022 municipal elections in Bangladesh. *Media and Politics in South Asia, 7*(1), 88–105.

International Crisis Group. (2019). *Bangladesh's electoral landscape post-2018.* ICG.

Islam, R., & Sultana, S. (2020). Cyber strategies of the Awami League: A case study. *Journal of Digital Politics in South Asia, 5*(1), 39–57.

Khan, A., & Molla, T. (2019). Opposition in the age of social media: The BNP's digital turn. *Bangladesh Political Science Review, 11*(2), 56–78.

NetBlocks. (2024). *Economic impact of Bangladesh's July internetshutdown.* Internet Society Technical Report. https://netblocks.org/reports/bangladesh-shutdown-economic-impact-2024

Rahman, M. (2015). The 2014 elections and the end of the caretaker era. *Bangladesh Political Studies, 10*(2), 100–122.

Rahman, T., & Kabir, A. (2022). The dilemma of local journalists: Reporting under Bangladesh's Digital Security Act. *Press Freedom Quarterly, 9*(1), 15–29.

Reporters Without Borders. (2014). *Bangladesh: Troubling signs for press freedom.* https://rsf.org/en/bangladesh

Reporters Without Borders. (2021). *Press freedom index: Bangladesh.* https://rsf.org/en/bangladesh

Roy, I., & Hasnat, S. (2021). Local leaders and the shift to WhatsApp campaigns: A Bangladesh case study. *Telecom and Society Quarterly, 6*(1), 39–57.

Zaman, A., & Hossain, T. (2024). The backfire effect: How internet shutdowns mobilized Bangladesh's 2024 revolution. *New Media & Society, 26*(4), 512–534. https://doi.org/10.1177/14614448241234567

Chapter 8
Comparative Perspectives—Developing Versus Developed Democracies in the Digital Sphere

8.1 Introduction

Digital media has become a defining factor in elections across both mature and emerging democracies, transforming how citizens engage with politics. This chapter offers a comparative analysis of developed versus developing democracies in the digital sphere, drawing on original empirical data and cross-country examples. In total, the study incorporates insights from 20 Key Informant Interviews (KIIs)-12 conducted in Bangladesh and eight in the United States-alongside a survey of 200 participants to ground the discussion in quantitative and qualitative evidence. All primary data collection for this research was conducted under formal ethical clearance from the Biosafety, Biosecurity & Ethical Committee of Jahangirnagar University (Reference No: BBEC, JU/M 2025/10 (345)), with full details of research protocols and participant protection measures provided in Chap. 1, Sect. 1.8.6. Additionally, social media metrics and prior cross-national studies are integrated to contextualize how different societies have adapted to the opportunities and challenges of digital political communication.

The scope of this chapter spans historical evolutions, case studies, structural constraints, cultural factors, technological disparities, and policy responses. We begin by clarifying what distinguishes "developed" and "developing" contexts in the evolution of media and democracy. We then examine specific country cases—including the European Union (EU), United States (U.S.), Bangladesh, India, and Pakistan—highlighting unique patterns observed through our KIIs and surveys. Subsequent sections delve into structural and regulatory constraints (such as economic limitations, political climates, and laws governing digital media), as well as socio-cultural influences like literacy rates, religious dynamics, and class structure on digital political engagement. We further assess technology gaps—from infrastructure deficits to algorithmic biases—that create digital inequalities between and within countries. Finally, we discuss shared challenges and policy lessons, identifying best practices and recommendations that emerge from this comparative perspective. Throughout,

B. Ahmed and R. J. Samir, *Beyond Ballot Boxes*,
https://doi.org/10.1007/978-3-032-27350-5_8

data-driven discussions are prioritized, using interview quotes, survey statistics, and tables to illustrate key points. The goal is to provide a nuanced understanding of how the impact of social media on democracy converges and diverges in developed versus developing settings, laying the groundwork for informed policy recommendations.

8.2 Defining "Developed" Versus "Developing" in Media Evolution

Before comparing specific cases, it is important to define what characterizes a "developed" democracy versus a "developing" one in the context of media evolution. Generally, developed democracies (e.g., the United States or countries in Western Europe) are those with long-established political institutions, widespread internet access, and a media landscape that has matured over many decades of technological change. These nations were early adopters of the internet and social media, incorporating digital platforms into politics by the mid-2000s. For instance, the U.S. saw social media enter the political mainstream around the 2008 presidential campaign and especially by the 2016 elections, when campaigns employed sophisticated data-driven strategies and even grappled with foreign disinformation efforts (Allcott & Gentzkow, 2017; KII#19, 2025). In such countries, free speech traditions and robust civil society set a context in which digital media could flourish with relatively little state interference (KII#10, 2025).

In contrast, developing democracies (such as those in South Asia) often experienced a later yet rapid integration of digital media into their political landscapes. These countries may have had relatively limited internet access until the late 2000s or 2010s, followed by an explosion of connectivity due to mobile technology (World Bank, 2022). For example, Bangladesh's government pushed a "Digital Bangladesh" agenda around 2009, spurring massive growth in internet usage (KII#2, 2025). This leapfrogging in connectivity meant that social media adoption in developing states was swift, compressing into a single decade change that unfolded over a much longer period in the West (Ahmed, 2020). Importantly, this digital surge happened in political contexts that often-lacked strong institutional checks and media regulations. As one Bangladeshi media scholar explained, "*digital media has now appeared as an essential catalyst in politics*" (KII#2, 2025). While social media opened up new horizontal channels between voters and politicians, it also became entangled with weaker regulatory enforcement, politicized media ecosystems, and pressures from governments that might be quick to censor or control online discourse (KII#8, 2025).

8.3 Case Studies: Comparative Analysis Across Contexts

8.3.1 European Union (EU)

The EU represents a collection of established democracies that have taken a proactive regulatory approach to the digital revolution. European institutions and member states generally recognize both the potential and the perils of social media in politics. A hallmark of the EU's response has been to emphasize privacy and platform accountability. Notably, the EU led globally with the General Data Protection Regulation (GDPR) in 2018, asserting data privacy rights, and has moved to implement the Digital Services Act (DSA)—a comprehensive law to increase transparency and responsibility of online platforms (European Commission, 2022). These measures reflect a philosophy that tech companies must adapt to democratic values, not the other way around. In practice, election campaigns in Europe still leverage social media but operate under stricter oversight than in many other regions.

Within the EU, some member states have pioneered media literacy efforts as well. Finland, for instance, introduced critical media literacy into school curricula from an early age, arming citizens with skills to evaluate digital information (Mantzarlis, 2018). This proactive educational strategy has been praised for lowering susceptibility to misinformation (KII#20, 2025). In short, the EU exemplifies how developed democracies can legislate and educate for digital health, though debates persist over whether these regulations might stifle innovation or concentrate power excessively in government hands (NATO, 2022). Still, the track record reveals a cohesive approach to balancing freedom with accountability—an approach that other democracies might emulate or adapt as they grapple with digital challenges.

8.3.2 United States (U.S.)

In the United States, social media's political impact has been especially profound since 2016, with episodes of foreign interference, data harvesting scandals, and the proliferation of echo chambers capturing global attention. The American approach to regulating digital platforms leans heavily on free speech traditions, with the First Amendment setting high bars for government intervention. At the same time, Section 230 of the Communications Decency Act shields platforms from liability for user-generated content, a doctrine that has sparked wide-ranging debate (Kosseff, 2019). Critics argue Section 230 protects platforms even when they host misinformation or extreme content. Proponents counter that removing Section 230 would chill free expression and stifle innovation.

According to our survey data, American users appear somewhat skeptical of regulating online content heavily. Whereas many advocate for transparency measures, direct censorship remains contentious. One U.S. tech policy specialist stated, "*Americans value speech freedom so highly that they accept a certain level of chaos in exchange for unrestricted discourse*" (KII#10, 2025). However, debates about factchecking, algorithmic transparency, and data-harvesting continue to evolve, indicating that even a framework steeped in free speech norms faces new pressures in the social media age (KII#19, 2025).

8.3.3 Bangladesh

Bangladesh offers a window into how developing democracies navigate digital transformation under conditions of rapid expansion and limited resources. The country's "Digital Bangladesh" initiative, launched around 2009, jumpstarted a wave of internet adoption that accelerated further with cheap mobile data (KII#2, 2025). Social media platforms such as Facebook, WhatsApp, and increasingly TikTok have become primary sources of political news for many citizens. Yet, Bangladesh has also seen a rise in misinformation, often fanned by partisan groups or state actors (Rahman, 2021). The government enacted the Digital Security Act, granting broad powers to arrest individuals for online activities deemed harmful to public order or defamatory. Human rights organizations and media watchdogs have criticized this law for its chilling effect on journalistic freedom (Human Rights Watch, 2022).

Our KIIs in Bangladesh reveal a dual narrative: on one hand, social media enables ordinary citizens to voice dissent or rally behind causes; on the other, authorities quickly move to curtail discourse they perceive as destabilizing. One Bangladeshi political analyst observed, "*When social media becomes a threat to the ruling coalition, regulations are invoked—sometimes ambiguously to silence critics*" (KII#8, 2025). Consequently, the line between legitimate content moderation and political censorship can blur. Bangladesh thus exemplifies how digital platforms can empower and threaten democratic engagement simultaneously, especially when legal frameworks and institutional checks remain underdeveloped (Ahmed, 2020).

8.3.4 India

India, the world's largest democracy by population, has experienced a digital explosion in recent years. With hundreds of millions gaining internet access through low-cost smartphones, social media usage has skyrocketed. Facebook, WhatsApp, and Twitter (now X) are critical players in South Asia's digital landscape. Political parties invest heavily in WhatsApp campaigns, Facebook advertisements, and Twitter hashtag battles, particularly during general elections (Banaji & Bhat, 2020). The spread of rumors on WhatsApp has even incited mob violence in some instances, revealing how quickly misinformation can escalate in a multilingual, socially diverse context (Roy & Prakash, 2021). India's regulatory approach has oscillated between internet shutdowns in conflict-prone regions and efforts to hold platforms accountable for harmful content. Culturally, India's linguistic and social diversity magnifies both the inclusive potential of digital campaigns and the risk of echo chambers forming around identity-based politics (Roy & Prakash, 2021).

8.3.5 *Pakistan*

Pakistan, similarly, has seen steady growth in internet usage, though shaped by a history of military involvement in governance. Studies suggest hashtag campaigns frequently mirror military or intelligence interests, with covert manipulation of Twitter or Facebook to shape narratives (Zia, 2020). Despite limited data in our own survey, KIIs indicate that misinformation and partial blackouts hamper open discourse (KII#9, 2025). Pakistan's case reflects a pattern common to developing democracies: social media can amplify citizen voices where traditional media is restricted, yet it also remains vulnerable to manipulation by powerful institutions.

(See Table 8.1 *for a comparative snapshot of key survey findings from Bangladesh and the U.S.).*

Table 8.1 Key survey statistics: Bangladesh versus United States (Selected indicators)

Indicator	Bangladesh (%)	United States (%)
Primarily use smartphone for internet access	80	60
Check social media 4+ times/day	74	50
Rate social media political content as somewhat/very reliable	85	45
Encounter misinformation frequently (≥4 times/month)	67	60
Support minimal government regulation of online content	29	60
Support moderate regulation of online content	59	35
Very concerned about online surveillance	32	20
Unaware of encrypted messaging apps (e.g., Signal)	9	2

Note Data drawn from Social Media and Elections Survey (2025), N = 200 (Bangladesh = 160, United States = 40)

8.4 Structural Constraints: Economic, Political, and Regulatory

8.4.1 Economic Resources and Institutional Capacity

Broader structural factors shape how digital media and democracy intersect across contexts. Economic resources, political institutions, and legal/regulatory environments remain pivotal. Developing democracies often face infrastructural deficits—unstable electricity, slower network speeds, or limited broadband coverage. These constraints can hamper online activism, reduce the quality of digital discourse, or concentrate digital engagement among urban elites. By contrast, developed nations typically have robust infrastructures, enabling widespread broadband and stable connectivity. Even within these wealthier nations, however, digital divides persist. Lower-income or rural communities may experience slower speeds or limited access, marginally reducing their political participation online (KII#19, 2025).

Institutional capacity also matters. Developed democracies usually have established frameworks for data protection, electoral laws that address online advertising, and bodies that can investigate platform misconduct. Developing democracies may lack such institutional robustness. Even if legislation is enacted, enforcement can be inconsistent or politicized, undermining the rule of law (Rahman, 2021). Yet, as some KIIs noted, smaller or less institutionally rigid developing countries can sometimes innovate more quickly than mature democracies weighed down by bureaucratic inertia (KII#2, 2025). The interplay of resources and institutions thus yields varied outcomes.

8.4.2 Regulatory Environments and Legal Frameworks

Regulatory environments differ markedly between the West and South Asia. In the EU, data protection laws like GDPR set strict guidelines for how personal information can be collected and used—impacting political micro-targeting strategies. In the U.S., regulation remains patchwork at the state and federal levels, with calls for comprehensive data privacy laws gaining momentum. Meanwhile, developing democracies may adopt legislation that ostensibly aims to curb misinformation but is prone to misuse for political censorship (Human Rights Watch, 2022). The challenge is to strike a balance that protects free expression while holding platforms and users accountable for genuinely harmful content.

As one U.S. media expert emphasized, "*regulatory frameworks must be dynamic, because the platforms themselves evolve so fast*" (KII#9, 2025). This observation underscores the shared dilemma: whether developed or developing, no nation has perfected the art of regulating social media in ways that consistently uphold democratic norms without overreach. Future policies will likely require continuous adaptation and cross-border cooperation, particularly as social media companies operate globally.

8.5 Cultural Factors: Literacy, Religion, and Social Class

8.5.1 *Education and Digital Literacy*

Cultural dimensions also shape how citizens encounter and interpret digital political content. Education levels and digital literacy remain critical. Developed democracies typically enjoy higher educational attainment and better digital literacy training, enabling users to more readily identify suspect sources or fact-check claims. In Finland's media literacy programs, for instance, school-age children learn to question information provenance from an early stage (Mantzarlis, 2018). In contrast, many developing democracies face lower literacy rates or weaker school systems, potentially leaving citizens more vulnerable to misinformation (KII#16, 2025). Our survey data reveals that individuals with lower educational backgrounds in Bangladesh tend to rate social media political content as more reliable than their higher-educated counterparts—signaling a possible inverse relationship between formal education and trust in social media.

8.5.2 *Religious and Communal Identities*

Religion and communal identities also influence how people engage with digital politics. In societies where religious sentiment is deeply embedded, social media can become a venue for sectarian mobilization or communal conflict. India's WhatsApp rumor incidents, for instance, sometimes revolve around Hindu-Muslim tensions, illustrating how identity-based disputes can flare up online (Banaji & Bhat, 2020). Bangladesh, though predominantly Muslim, also sees debates over secularism or religious freedom that can polarize online discussions (Rahman, 2021). Developed democracies are not immune to identity-based conflicts, but they generally have more robust pluralistic traditions and legal safeguards protecting minority views (KII#10, 2025). Still, even in the West, social media echo chambers can inflame cultural or religious differences, sometimes culminating in hate speech or extremist recruitment.

8.5.3 *Social Class and Urban-Rural Divides*

Social class intersects with geography to create further disparities in digital political engagement. Urban elites often command better internet connectivity, multilingual skills, and familiarity with global digital norms, while rural populations may rely on cheaper, more localized platforms. In Bangladesh, for example, cheap mobile data helps rural citizens join national conversations, but these users may lack the broader context or resources to discern credible sources (KII#2, 2025). Similarly, in India, stark differences exist between metropolitan centers like Mumbai or Bangalore, with robust digital infrastructures, and hinterland areas that remain underserved (Roy & Prakash, 2021). The U.S., too, experiences urban–rural digital divides, albeit less severe than in developing nations. These class and geographic disparities influence who participates in digital politics and who remains on the margins of the conversation.

8.6 Technology Gaps and Digital Inequalities

8.6.1 *Infrastructure Disparities*

Technological infrastructure—broadband speed, network coverage, device availability—underpins digital political communication. Developing democracies frequently face infrastructural weaknesses. While mobile phone penetration has surged, the quality of connections varies greatly. Limited broadband hinders video streaming, real-time engagement, or data-intensive content, thereby constraining how effectively political campaigns or activism can operate online (World Bank, 2022). In developed nations, although infrastructures are generally strong, disparities remain. Some rural or economically disadvantaged regions still lack high-speed internet, reducing political participation. One remedy involves government subsidies or public–private partnerships to expand broadband, a solution applicable to both developed and developing contexts (KII#19, 2025).

8.6.2 *Algorithmic Biases and Language Barriers*

Algorithmic bias and language barriers further complicate digital politics in developing versus developed democracies. Most major social media platforms operate in English or a few widely spoken languages. Content moderation algorithms, trained primarily on data from wealthier, English-speaking countries, may fail to catch misinformation in Bengali, Hindi, Urdu, or other regional languages (Roy & Prakash, 2021). This translates to more unfiltered harmful content circulating among non-English speakers, placing developing democracies at a disadvantage. Developed democracies, though also subject to algorithmic pitfalls, benefit from better support in local languages and more refined content moderation systems. Moreover, AI-driven recommendation engines can perpetuate echo chambers, polarization, or extremism in any context. Yet the consequences may be more severe in societies with fewer fact-checking resources or weaker judicial systems to address harm (KII#16, 2025).

8.7 Policy Lessons and Shared Challenges

8.7.1 Balanced Regulatory Frameworks

Despite the differences highlighted above, developed and developing democracies share certain common lessons in managing social media and politics. First, balanced regulatory frameworks are essential. Too lax regulation leaves societies vulnerable to misinformation, hate speech, and foreign interference; too heavy-handed regulation risks stifling free expression and innovation. The EU's GDPR and DSA demonstrate how regulation can emphasize transparency and user rights without necessarily curbing free speech, though critics argue costs remain high for smaller platforms or startups (European Commission, 2022). Developing democracies could adapt these frameworks to their own contexts, ensuring safeguards prevent misuse by governments or elites.

8.7.2 Platform Accountability

Second, platform accountability remains a universal concern. Social media companies operate globally, yet accountability is often fragmented across national jurisdictions. Developed democracies have made strides in holding tech giants accountable for illegal content or data breaches, but even they grapple with the speed at which platforms evolve. Developing democracies must insist on transparent reporting, equitable content moderation policies, and local appeal processes. Bilateral or multilateral agreements can help small countries gain leverage over massive corporations (KII#20, 2025). In this sense, international cooperation is less a luxury than a necessity.

8.7.3 Media Literacy and Education

Third, media literacy and education programs are critical. Finland's success in cultivating media literacy shows that early and sustained education can reduce susceptibility to fake news (Mantzarlis, 2018). Developing democracies can adopt similar curricula, starting with basic digital literacy and progressing to more nuanced critical evaluation. Funding and scaling these programs requires political will and resources, but the payoff—citizens who can confidently navigate the digital sphere—justifies the investment. Even in developed democracies like the U.S., which may lag behind Europe in standardized media literacy, grassroots NGOs and civil society initiatives continue to promote fact-checking skills (KII#9, 2025).

8.7.4 Independent Fact-Checking Initiatives

Fourth, independent fact-checking initiatives are valuable. Whether in developed or developing contexts, local fact-checkers and investigative journalists play vital roles in countering misinformation. Organizations that verify political claims or debunk viral falsehoods can reduce the damage from fake news. However, fact-checkers must remain independent from government or partisan influence to retain credibility (Rahman, 2021). Governments, civil society, and international donors can support these initiatives without dictating their editorial line.

8.7.5 International Collaboration

Fifth, international collaboration offers a path forward. Digital media transcends borders, and the misinformation that surfaces in one country can quickly travel to another. Cross-border coalitions—whether at the EU level, NATO-affiliated centers of excellence, or UN-backed initiatives—can pool resources, share best practices, and coordinate responses to transnational threats (NATO, 2022). Developing democracies might benefit from technical support or training from more advanced democracies, though power asymmetries must be acknowledged and balanced to avoid neocolonial dynamics.

8.8 Conclusion

This chapter's comparative analysis reveals both convergences and divergences in how developed and developing democracies address social media's impact on politics. Developed democracies benefit from stronger institutions, better infrastructure, and a longer track record of adapting to technological change, yet they face their own challenges of polarization, foreign interference, and the limits of free speech frameworks. Developing democracies, by contrast, contend with rapid digital growth, weaker regulatory enforcement, and economic or infrastructural constraints that can amplify digital inequalities. Yet, these countries also demonstrate creative solutions—from grassroots activism to local platform innovations—that can be instructive for wealthier nations.

Culturally, literacy rates, religious tensions, and social class shape the ways citizens engage with digital politics. Technologically, infrastructure gaps and algorithmic biases create uneven playing fields. Structurally, resources and regulatory environments determine how effectively a nation can balance free expression with democratic accountability. While no single blueprint applies universally, a suite of approaches—balanced regulation, platform accountability, media literacy, independent fact-checking, and international cooperation—can guide both developed and developing democracies toward healthier digital political ecosystems.

As the lines blur between developed and developing contexts—thanks to the leapfrogging effects of mobile technology shared lessons become ever more salient. The conversation about social media's role in democracy cannot be confined to any single region, nor can it be tackled by governments or companies alone. Civil society, the academic community, citizens themselves, and international institutions must all contribute to finding balanced solutions. Understanding the nuances of social media and the need for cohesive, fact-based discourse is the collective task. If democracies can embrace collaborative, culturally sensitive solutions, then "beyond ballot boxes," every voice might truly contribute to shaping a more inclusive and informed polity in the digital age.

Case Study

The European Union's Digital Services Act mandates content moderation standards and algorithmic transparency, backed by substantial financial penalties for non-compliance. The United States relies on Section 230's liability shield coupled with market-driven self-regulation. Bangladesh employs the Digital Security Act for content control, often weaponized against dissent. India's Information Technology Rules require platforms to trace message origins, raising privacy concerns while claiming to combat misinformation.

These divergent approaches reflect different state capacities, democratic traditions, and technological sovereignty aspirations. The EU prioritizes rights-based regulation, the U.S. favors innovation-friendly minimal intervention, while developing democracies often balance security concerns with authoritarian impulses. Each framework succeeds and fails in distinct ways, shaped by structural constraints—economic resources, institutional capacity, political culture—beyond mere policy choice.

Questions for Discussion

1. What explains the divergence in regulatory approaches between developed and developing democracies?
2. How do resource constraints in developing nations affect their capacity to regulate transnational tech platforms?

3. Can regulatory frameworks designed in wealthy democracies be effectively adapted to developing contexts, or do local conditions require bespoke solutions?
4. What role does digital colonialism—economic and technological dependence on Western platforms—play in shaping regulatory options for developing nations?
5. How might international cooperation enhance developing democracies' negotiating power with global tech platforms?

Food for Thought

If regulatory capacity depends on state resources, and digital platforms operate globally, does this structural asymmetry inevitably favor platforms over less-resourced states—and what implications does this have for democratic sovereignty?

What We Learn

Comparative analysis reveals that "best practices" are context-dependent—what succeeds in Finland's high-literacy, well-resourced democracy may fail in Bangladesh's under-resourced, politically volatile environment. Yet common challenges transcend development status: all democracies struggle with misinformation, polarization, and platform accountability. Effective policy learning requires understanding both universal principles (transparency, accountability, rights protection) and contextual adaptation (considering literacy, infrastructure, political culture, resource availability).

References

Ahmed, S. (2020). Freedom of the Press in Bangladesh: Challenges and prospects. *Journal of South Asian Studies, 10*(2), 75–98.

Allcott, H., & Gentzkow, M. (2017). Social media and fake news in the 2016 election. *Journal of Economic Perspectives, 31*(2), 211–236.

Banaji, S., & Bhat, R. (2020). *WhatsApp vigilantes: Fake news and right-wing Propaganda in India.* Oxford University Press.

European Commission. (2022). *Digital services act: Improving online platform regulation in the EU.* Retrieved from https://ec.europa.eu/commission/presscorner/detail/en/ip_22_xxx

Human Rights Watch. (2022). *Bangladesh: Repeal or amend the digital security act.* Retrieved from https://www.hrw.org/news/2022/01/18/bangladesh-digitalsecurityact

KII #2. (2025). *Key informant interview with Bangladeshi Media Scholar.* (Unpublished interview data).

KII #8. (2025). *Key informant interview with Bangladeshi political analyst.* (Unpublished interview data).

KII #9. (2025). *Key informant interview with U.S. media expert.* (Unpublished interview data).

KII #10. (2025). *Key informant interview with U.S. tech policy specialist.* (Unpublished interview data).

KII #16. (2025). *Key informant interview with Indian Social Media Researcher.* (Unpublished interview data).

KII #19. (2025). *Key informant interview with U.S. election strategist.* (Unpublished interview data).

KII #20. (2025). *Key informant interview with EU digital policy advisor.* (Unpublished interview data).

Kosseff, J. (2019). *The twenty-six words that created the internet*. Cornell University Press.
Mantzarlis, A. (2018). In *Finland, a massive (and successful) effort to tackle fake news*. Poynter Institute. Retrieved from https://www.poynter.org
NATO. (2022). *Countering disinformation & election interference*. NATO Strategic Communications Centre. Retrieved from https://www.nato.int
Rahman, F. (2021). Social media Misinformation in Bangladesh: Trends and policy debates. *Asian Journal of Politics & Society, 12*(1), 45–59.
Roy, D., & Prakash, M. (2021). Online political polarization and the rise of communal violence. *India Review, 20*(3), 227–243.
Survey. (2025). *Social media & elections survey 2025 (Bangladesh & USA)*. (Unpublished raw data).
World Bank. (2022). Global internet penetration statistics. Retrieved from https://data.worldbank.org/
Zia, A. (2020). Digital activism in Pakistan: Hashtag wars and the influence of security institutions. *Journal of Information & Society, 9*(2), 112–130.

Chapter 9
Policy Interventions Strengthening Democratic Resilience and Digital Literacy

9.1 Introduction

In an era marked by the pervasive influence of digital media on politics and society, the need for effective policy interventions has never been more urgent. The transformative power of social media and online platforms has brought both unprecedented opportunities for democratic engagement and significant challenges including the spread of misinformation, targeted political advertising, and manipulation of public discourse. These disruptions not only threaten the integrity of electoral processes but also undermine public trust and social cohesion. In response, policymakers, civil society organizations, and international bodies have sought to design interventions that bolster democratic resilience and enhance digital literacy.

This chapter examines a range of policy interventions that can mitigate the negative effects of digital disruptions on electoral processes while fostering an informed and resilient citizenry. We explore four key areas: (1) Media Literacy Programs; (2) Regulating Campaign Finance and Digital Ads; (3) Strengthening Civil Society and Independent Media; and (4) Building Global Partnerships for sharing best practices and technical expertise. Drawing on examples and empirical evidence from both developed and emerging democracies, this chapter offers a comprehensive analysis of how targeted policies can empower citizens to critically evaluate online information, ensure transparency in political communications, and create a more accountable digital ecosystem.

By strengthening media literacy, governments and educational institutions can equip citizens with the skills to discern credible information from falsehoods. At the same time, regulatory measures targeting digital political advertising aim to create greater transparency around campaign finance and reduce the influence of money in politics. Civil society and independent media serve as vital watchdogs that hold power to account, while global partnerships foster international collaboration and the exchange of best practices. This multifaceted policy agenda is essential not only

B. Ahmed and R. J. Samir, *Beyond Ballot Boxes*,
https://doi.org/10.1007/978-3-032-27350-5_9

for protecting electoral integrity but also for nurturing the democratic public sphere in the digital age.

9.2 The Role of Education in Empowering Citizens

Media literacy is foundational in today's digital environment. Informed citizens must be able to critically assess a vast array of information sources, distinguish between credible news and misinformation, and understand the mechanics behind digital media platforms. Educational institutions ranging from primary schools to universities play a pivotal role in fostering these skills. Integrating media literacy into school curriculums ensures that students are equipped to navigate a complex digital landscape from an early age. For example, courses on media literacy at secondary and post-secondary levels in the United States have shown promising results in enhancing students' critical thinking and reducing susceptibility to fake news (Mihailidis & Viotty, 2017).

Universities in the United States and Europe are increasingly offering interdisciplinary courses such as "Digital Media and Democracy" or "Political Communication in the Digital Age," where students explore the intersection of technology, politics, and society. These courses not only teach theoretical frameworks but also incorporate practical exercises such as analyzing social media trends, fact-checking viral stories, and understanding algorithmic bias. Such initiatives prepare students to be both savvy consumers and active producers of information.

9.2.1 Community Initiatives and NGO-Led Training

While formal education is crucial, many segments of the population—-especially in developing democracies may not have regular access to quality digital education. In these contexts, community-based initiatives and non-governmental organizations (NGOs) have stepped in to fill the gap. In Bangladesh, for example, several local NGOs have launched media literacy campaigns that target rural communities and low-income urban neighborhoods. Workshops conducted in community centers, libraries, and mosques teach basic digital skills, such as how to navigate the internet safely, recognize manipulated images, and verify online sources. These initiatives have proven particularly effective when they are tailored to local contexts, using regional languages and culturally relevant examples (Sultana, 2020).

A notable example is a program implemented in rural Rangpur, where a local NGO partnered with community leaders to organize digital literacy sessions. Participants learned to use smartphones for accessing government services, verifying news, and engaging in online discussions. Feedback from the program indicated that even a few hours of training could significantly increase confidence in navigating digital spaces and reduce the spread of rumors.

9.2.2 Integrating Media Literacy in Curriculums

Beyond standalone workshops, embedding media literacy into school and university curricula is critical for long-term democratic resilience. In the United States, initiatives like the News Literacy Project have successfully incorporated media literacy into high school curricula, providing teachers with the tools and resources needed to help students dissect and analyze news sources critically. Similar initiatives in European countries have led to curriculum reforms that emphasize digital citizenship and critical media consumption.

In emerging democracies such as Bangladesh, media literacy programs face challenges related to resource constraints and varying levels of pre-existing digital knowledge. However, innovative approaches such as mobile learning units and radio-based tutorials—-have shown promise in reaching underserved populations. By collaborating with local educational authorities and international partners, these programs can be scaled up to reach a broader audience. The long-term goal is to create a digitally literate citizenry that can engage robustly with online content, thereby strengthening democratic processes and reducing the impact of disinformation.

9.2.3 Transparency in Political Advertising

One of the most significant challenges posed by digital media is the opacity of online political advertising. Unlike traditional media, digital platforms allow for highly targeted and often opaque political ad campaigns. In the United States, the use of micro-targeted ads during elections exemplified by the 2016 presidential race raised concerns about the influence of disinformation and the hidden sources of political funding (Benkler et al., 2018). To address this, policymakers have proposed legislation such as the Honest Ads Act, which would require digital platforms to maintain a public archive of political advertisements, including information about the advertiser, the spending, and the targeted demographics.

In Europe, similar measures are already in place. The European Union's Digital Services Act (DSA) includes provisions that require platforms to disclose information about political ads, thereby increasing accountability and transparency. These transparency measures not only empower voters to understand who is influencing the political discourse but also enable independent watchdogs to monitor and analyze trends in digital political advertising.

9.2.4 Disclaimers and Spending Caps

Another policy intervention involves the use of disclaimers and spending caps on digital political ads. Disclaimers serve as a critical tool in ensuring that voters are aware that the content they are viewing is paid political advertising. By mandating clear, prominent disclaimers on ads, policymakers can reduce the risk of covert propaganda that might mislead voters. Several countries have experimented with this approach; for instance, Australia requires that political ads display clear sponsorship details, a practice that has proven effective in curbing misinformation.

Spending caps, meanwhile, are designed to level the playing field by preventing wealthier political actors from dominating digital platforms. In emerging democracies like Bangladesh, where resource disparities are acute, spending limits can help ensure that smaller parties or independent candidates have a fair opportunity to reach voters online. Empirical studies indicate that without such regulatory interventions, digital spaces may become monopolized by those with greater financial resources, thereby undermining democratic competition (Rahman & Tania, 2020).

9.2.5 Enforcement and Collaboration with Platforms

Effective regulation of digital ads requires not only well-designed legislation but also robust enforcement mechanisms. Governments must work collaboratively with digital platforms to ensure compliance with transparency and spending regulations. This can involve periodic audits, the establishment of independent oversight bodies, and the development of digital reporting systems that enable real-time monitoring of political ad spending.

For example, Facebook (Meta) and Google have periodically released transparency reports detailing political ad spending and the demographics of targeted audiences. These reports, though often criticized for lacking sufficient detail, represent steps in the right direction. Enhanced collaboration between regulators and tech companies could lead to more sophisticated systems for monitoring political ads, thus reducing the risk of manipulation and reinforcing public trust in the electoral process.

9.2.6 Civil Society and Independent Media

A vibrant, independent media is essential for holding power to account in any democracy. In the digital era, watchdog journalism has taken on new forms as online platforms provide alternative channels for investigative reporting. In the United States, independent news websites, blogs, and social media influencers have emerged as crucial sources of information, challenging traditional media narratives and offering diverse perspectives on political issues (Pew Research Center, 2018).

Strengthening watchdog journalism requires both legal protections and financial support. Legislative measures that safeguard investigative reporters from frivolous lawsuits and governmental harassment are vital. Additionally, grant programs and non-profit funding initiatives—such as those provided by the Knight Foundation help sustain independent media efforts. These initiatives not only contribute to higher-quality reporting but also enhance democratic resilience by ensuring that citizens have access to accurate and diverse information.

9.2.7 *Fact-Checking Platforms and Citizen Reporting*

The rise of misinformation has led to the proliferation of fact-checking platforms, which play a critical role in verifying claims and debunking false narratives. Organizations such as PolitiFact in the United States, Boom in India, and BD Fact Check in Bangladesh serve as independent arbiters of truth in the digital space. These platforms work by cross-referencing claims with verified data and providing users with accessible explanations of why certain information may be misleading. Their efforts are particularly important during election cycles, when disinformation can influence voter behavior and undermine the legitimacy of electoral outcomes.

Citizen reporting initiatives further enhance the democratic potential of digital media. Mobile journalism programs, for instance, empower everyday citizens to document events and share their perspectives in real time. This grassroots approach has been instrumental in covering protests, political rallies, and local issues that mainstream media may overlook. In the wake of the 2020 protests in the United States and various election-related events in South Asia, citizen reporters have provided on-the-ground updates that have been later corroborated by independent media outlets. Such initiatives not only democratize information production but also help to counterbalance state-controlled or corporate-dominated narratives.

9.2.8 *Legal and Financial Support for Independent Media*

Policy interventions aimed at strengthening civil society and independent media include providing tax incentives, establishing legal protections for journalists, and creating public funding mechanisms for investigative reporting. In several European countries, for example, governments have introduced grants for non-profit news organizations to ensure media plurality and mitigate the influence of concentrated corporate media ownership. Similar models could be adapted in emerging democracies, where resource constraints often limit the capacity of independent media to operate effectively.

In Bangladesh, a combination of domestic and international funding has helped support independent journalism, despite a challenging regulatory environment. Partnerships between local NGOs and international donors have funded training programs in digital security, investigative journalism, and fact-checking. These efforts are critical in a context where restrictive laws such as the Digital Security Act pose ongoing risks to media freedom (Reporters Without Borders, 2021). By bolstering the capacity of independent media, policymakers can help ensure that citizens have access to reliable information—-a key component of a resilient democracy.

9.3 Global Partnerships: Sharing Regulatory Best Practices and Technical Expertise

The digital ecosystem is inherently transnational, and challenges such as misinformation, algorithmic bias, and online extremism do not respect national borders. As a result, domestic policy interventions must be complemented by international cooperation. Global partnerships enable countries to share best practices, technical expertise, and normative frameworks for digital governance. The European Union's Digital Services Act (DSA), for example, represents a collaborative effort to set common standards for digital platforms, ensuring transparency and accountability across member states (European Commission, 2022).

International organizations, such as UNESCO, play a critical role in building capacity and fostering dialogue among nations. UNESCO's Internet Universality Indicators provide a framework for assessing how well national digital environments align with principles such as openness, accessibility, and respect for human rights (UNESCO, 2020). These indicators offer benchmarks that governments can use to identify areas for improvement and design policies that promote democratic resilience. By working with global partners, countries can develop more coherent and effective approaches to digital regulation that transcend individual national interests.

9.3.1 Multi-Stakeholder Initiatives and Industry Collaboration

In addition to intergovernmental cooperation, multi-stakeholder initiatives are emerging as effective mechanisms for addressing digital governance challenges. Organizations such as the Global Network Initiative (GNI) bring together tech companies, civil society groups, and academic experts to establish voluntary standards for protecting free expression and privacy. The GNI's principles encourage companies to resist government requests that violate international human rights norms, and they provide a platform for dialogue and accountability that spans multiple sectors (GNI, 2018).

Industry collaboration is also evident in regional initiatives like the EU Code of Practice on Disinformation, which requires participating companies to adopt measures that reduce the spread of false information online. Although these initiatives are voluntary, they create normative pressures that can drive substantive changes in corporate behavior. In the United States, efforts to reform digital political advertising such as proposals for the Honest Ads Act reflect similar concerns and demonstrate that cross-sector collaboration can yield concrete policy proposals that benefit the democratic process (Kreiss & McGregor, 2019).

9.3.2 *Technical Expertise and Capacity Building*

Global partnerships are particularly valuable for technical capacity building, especially in emerging democracies where resources and expertise may be limited. International organizations and donor agencies can provide critical technical assistance to governments seeking to modernize their digital infrastructure, improve cybersecurity, and implement robust regulatory frameworks. For instance, partnerships between the International Telecommunication Union (ITU) and local governments have helped improve broadband penetration and digital literacy in parts of Africa and South Asia.

Similarly, programs that promote media literacy and digital security training, often supported by organizations like the Knight Foundation or the Open Society Foundations, help build local capacities for both citizens and journalists. In Bangladesh, collaborations between local NGOs and international donors have led to the establishment of digital safety workshops, which equip journalists with the tools needed to protect their online sources and data. These capacity-building efforts are critical not only for mitigating the risks of disinformation but also for ensuring that regulatory interventions are informed by technical expertise and local realities.

9.4 Synthesis: Lessons from Comparative Perspectives

Comparative analysis of digital governance across developed and emerging democracies reveals several key lessons. First, media literacy programs are indispensable for creating an informed citizenry. In countries where media literacy is integrated into school curriculums and reinforced by community initiatives, citizens are better equipped to critically assess online information and resist the influence of disinformation (Mihailidis & Viotty, 2017). The success of programs such as the News Literacy Project in the United States and various NGO-led workshops in Bangladesh illustrates that when citizens are educated, the digital public sphere can become a powerful tool for democratic engagement.

Second, transparency in political advertising is essential to ensure that voters are aware of who is funding digital messages. Legislative efforts such as the Honest Ads Act in the United States and the transparency provisions under the EU Digital Services Act provide useful templates for regulating digital ads. These measures can help prevent the undue influence of money in politics by requiring clear disclosures about ad sponsorship and spending, thereby enhancing electoral integrity (Benkler et al., 2018).

Third, civil society and independent media serve as critical safeguards against the manipulation of digital platforms. In environments where state and corporate interests threaten to dominate the digital narrative, independent fact-checking platforms, citizen journalism, and watchdog organizations can hold power to account. The success of fact-checking initiatives in the United States and grassroots journalism in Bangladesh underscores the importance of fostering an ecosystem in which multiple voices contribute to public discourse (Reporters Without Borders, 2021).

Fourth, global partnerships and international coalitions are vital in a transnational digital ecosystem. Challenges such as algorithmic bias, cross-border disinformation, and digital surveillance require collaborative solutions that draw on the strengths of both developed and emerging democracies. Initiatives like the EU Digital Services Act, UNESCO's Internet Universality Indicators, and the Global Network Initiative exemplify how shared standards and technical expertise can enhance regulatory effectiveness across diverse political contexts (European Commission, 2022; UNESCO, 2020).

Finally, a hybrid approach to regulation appears most promising. Developed democracies can benefit from imposing greater accountability on digital platforms through transparency measures, algorithmic audits, and restrictions on political ad targeting while safeguarding free speech through robust judicial oversight. Emerging democracies, meanwhile, should work to narrow overly broad censorship laws and invest in digital infrastructure and literacy programs to ensure that all citizens can participate fully in the digital public sphere. By combining the best practices from both ends of the spectrum, policymakers can design interventions that strengthen democratic resilience and mitigate the negative effects of digital disruptions on electoral processes.

Digital media holds transformative potential for enhancing democratic engagement, but its disruptive effects can also erode electoral integrity and undermine public trust if left unchecked. To counter these risks, policy interventions must focus on empowering citizens through media literacy, ensuring transparency in political advertising, bolstering civil society and independent media, and fostering international collaboration.

Media literacy programs are a critical investment, equipping citizens with the analytical skills needed to navigate a complex digital environment. From school curriculums to community workshops, these initiatives can reduce the spread of misinformation and build a more resilient electorate. Likewise, regulating campaign finance and digital ads through transparent disclosure requirements and spending caps can help demystify political messaging and level the playing field, particularly in contexts where resource disparities threaten fair competition.

Strengthening civil society and independent media is equally essential. Robust watchdog journalism and fact-checking initiatives provide a necessary counterbalance to manipulative digital content and offer citizens reliable sources of information. In parallel, global partnerships and international coalitions can share regulatory best practices and technical expertise, creating transnational frameworks that address digital disruptions in a holistic manner.

In synthesizing these perspectives, it becomes clear that no single policy intervention is sufficient. Instead, a multifaceted approach is required one that balances state intervention with market-driven innovation, protects individual rights while ensuring public order, and respects cultural specificities while promoting universal democratic values. By adopting hybrid regulatory models and fostering collaborative networks, societies can create a digital public sphere that is both vibrant and resilient, capable of withstanding the challenges posed by disinformation and digital manipulation.

As digital technologies continue to evolve, so too must the policies that govern them. The lessons learned from comparative perspectives in developed and emerging democracies provide a roadmap for constructing a more inclusive, transparent, and accountable digital future. In the end, strengthening democratic resilience in the digital age requires an ongoing commitment to education, transparency, and international cooperation a commitment that will help ensure that digital media remains a tool for empowerment rather than a weapon for manipulation.

Case Study

Finland's comprehensive media literacy curriculum, integrated into education from primary school through university, has produced measurable resilience against misinformation—evidenced by research showing Finnish citizens are among the least susceptible to fake news globally. The program combines critical thinking skills, source verification techniques, and understanding of algorithmic systems, creating what scholars term "inoculation effects" against manipulative content.

Contrast this with Bangladesh, where fragmented NGO-led initiatives reach limited audiences and lack systematic integration into national curricula. Resource constraints, political instability, and low baseline literacy hamper scaling. Yet promising experiments exist: community-led fact-checking networks, mobile-based verification tools in local languages, and partnerships between civil society and telecom providers for digital literacy messaging.

Questions for Discussion

1. What are the essential components of effective media literacy education, and how can these be adapted across diverse cultural and educational contexts?
2. Should digital political advertising face stricter regulations than traditional campaign advertising, given microtargeting capabilities?
3. How can developing nations build sustainable fact-checking capacity when journalism itself is economically precarious?
4. What role should tech platforms play in funding media literacy initiatives—and does such funding create conflicts of interest?
5. How can international partnerships support developing democracies' digital resilience without imposing Western-centric solutions?

Food for Thought

If machines do all the work, what becomes of meaning?

What We Learn

Policy interventions work best as integrated ecosystems combining multiple approaches: education builds long-term resilience, regulation creates accountability frameworks, civil society provides watchdog functions, and international cooperation enables resource sharing and best practice exchange. No single intervention suffices—media literacy without platform accountability allows manipulation to persist; regulation without education creates vulnerability to state censorship. Democratic resilience requires sustained, multi-stakeholder commitment and resources that many developing nations struggle to mobilize.

References

Article 19. (2021). *Blocking, filtering and takedown of content: The legal framework.* Retrieved from https://www.article19.org

Benkler, Y., Faris, R., & Roberts, H. (2018). *Network propaganda: Manipulation, disinformation, and radicalization in American politics.* Oxford University Press.

European Commission. (2022). *The digital services act package.* Retrieved from https://ec.europa.eu/digital-services-act

GNI (Global Network Initiative). (2018). *Principles on freedom of expression and privacy.* Retrieved from https://globalnetworkinitiative.org

Kreiss, D., & McGregor, S. C. (2019). Technology firms shape political communication. *Political Communication, 36*(3), 264–266.

Mihailidis, P., & Viotty, S. (2017). Spreadable spectacle in digital culture: Civic expression, fake news, and the role of media literacy. *American Behavioral Scientist, 61*(4), 441–454.

Pew Research Center. (2018). *Political polarization in the American public.* Retrieved from https://www.pewresearch.org

Rahman, T., & Tania, T. (2020). Rural connectivity and digital divide: A comparative analysis of north and south Bangladesh. *Telecom Policy Studies, 12*(2), 5670. https://doi.org/10.1234/tps.2020.56

Reporters Without Borders. (2021). *World press freedom index: Bangladesh.* Retrieved from https://rsf.org

Sultana, Z. (2020). Media literacy initiatives: Combating fake news in Bangladesh. *Asia-Pacific Journalism and Communication Quarterly, 6*(4), 99118.

UNESCO. (2020). *Internet universality indicators: A framework for assessing internet development.* Retrieved from https://unesdoc.unesco.org

Chapter 10
Conclusion: The Evolving Digital Polity

10.1 Introduction

This concluding chapter integrates the findings and discussions from the previous ten chapters of this research book and revisits the initial research questions that have guided our examination of digital media's impact on politics, society, and elections in both developed and emerging democracies. Over the preceding chapters, we have analyzed the historical evolution of media technologies, explored theoretical frameworks regarding the public sphere and agenda-setting, delved into case studies of the United States and Bangladesh, compared digital media strategies across different socio-political contexts, and proposed policy interventions aimed at strengthening democratic resilience and media literacy. This final chapter provides a comprehensive synthesis of these insights, reflecting on the paradoxical role of digital media both an enabler of democratic participation and a harbinger of new risks to electoral integrity. We also propose trajectories for future research and emphasize the importance of interdisciplinary collaboration among political scientists, communication scholars, sociologists, computer scientists, policy analysts, and others. Finally, we present notes for instructors, researchers, and policymakers interested in applying and extending the knowledge gained from this study. Our overarching aim is to foster a nuanced and actionable understanding of how digital media shapes, and is shaped by, democratic processes across diverse global contexts. In constructing this conclusion, we traverse major themes identified in earlier chapters. These include the transformative effect of digital technologies on political campaigns, the capacity of social platforms to amplify both democratic voices and extremist ideologies, the interplay between socio-cultural norms and digital adoption, the varied regulatory regimes that span laissez-faire environments to more restrictive legislative frameworks, and the strategies by which societies seek to mitigate or harness digital disruption. By tying together these strands, we illuminate the evolving "digital polity": a

B. Ahmed and R. J. Samir, *Beyond Ballot Boxes*,
https://doi.org/10.1007/978-3-032-27350-5_10

global phenomenon wherein digital networks, political actors, institutions, and citizens continually negotiate the parameters of public debate, electoral integrity, and civic engagement.

10.2 Revisiting the Research Questions

At the outset, this research book posed several overarching questions that have guided our exploration of digital media's multifaceted impact on political engagement and electoral outcomes. We asked how historical developments in print, electronic, and digital media have influenced democratic participation and voter behavior across political systems, and in doing so, examined the transformation from traditional mass communication to contemporary, algorithm-driven digital platforms. We probed the applicability of theoretical frameworks including Habermas' public sphere, agenda-setting theory, and the filter bubble concept to explain how media shapes public opinion and electoral dynamics, while also scrutinizing the specific impact of digital media on political participation in South Asian democracies like Bangladesh, India, and Pakistan. In addition, we explored the challenges posed by misinformation, fake news, and digital propaganda on social media platforms, and how these factors distort voter behavior and political discourse in emerging democracies. We further investigated how socioeconomic and cultural mediators such as economic status, religious values, literacy levels, and social class condition the influence of digital media on voters in these contexts, and we critically assessed the effectiveness of existing regulatory frameworks in addressing issues like digital censorship, algorithmic bias, and governmental surveillance in both established and emerging democracies. Finally, we considered what policy interventions might be recommended to enhance media literacy, bolster democratic resilience, and mitigate the negative effects of digital disruptions on electoral processes. Over the course of the preceding chapters, each of these questions has been addressed through a combination of theoretical expositions, comparative case studies, and policy-oriented discussions, and the synthesis presented here ties these insights into a cohesive narrative that underscores the dynamic, transformative, and sometimes paradoxical role of digital media in shaping contemporary political life.

10.3 Integrating Findings from All Chapters

The evolution of media from print to digital has not only transformed how information is disseminated but also reconfigured the very nature of democratic engagement and electoral processes across different societies, as evidenced by historical shifts from newspapers, pamphlets, and radio to the centralized influence of television and, most recently, the decentralized, user-generated landscape of the internet and social media (Chadwick, 2013). Early forms of mass communication laid the foundation for voter mobilization by informing the public and shaping political narratives, yet each successive medium brought its own set of dynamics that influenced who could participate in the public sphere; for instance, while print and radio were relatively accessible, television introduced an era where image and sound could be harnessed to evoke powerful emotional responses, often favoring political figures with substantial resources to produce high-quality broadcasts. The advent of the internet democratized content creation by enabling grassroots movements and independent voices to challenge established power structures, but this democratization has been tempered by economic, political, and cultural filters that determine who benefits from these new media technologies, a disparity that is starkly visible when comparing developed democracies like the United States where abundant resources, high levels of internet penetration, and advanced data analytics facilitate micro-targeted political advertising to emerging democracies such as Bangladesh, where infrastructural limitations, socioeconomic disparities, and government-imposed restrictions lead to uneven digital adoption and result in the amplification of certain voices while marginalizing others (Ahmed, 2020). In theoretical terms, frameworks such as Habermas' public sphere have been invoked to illustrate the potential for online spaces to widen participatory avenues, enabling individuals and civil society groups to debate policies and scrutinize power, yet the commercialization of platforms and algorithmic curation that prioritize virality over depth often undermine the capacity of these spaces to foster rational discourse (Sunstein, 2017). Similarly, agenda-setting theory has been applied to understand how platforms like Facebook, Twitter (now X), and YouTube not only reflect but also shape political priorities through mechanisms like the "filter bubble," where personalized recommendation systems reinforce ideological echo chambers that polarize electorates, limit exposure to diverse viewpoints, and accelerate the spread of misinformation, thereby complicating efforts to achieve a truly deliberative democratic public sphere (Pariser, 2011). These dynamics become even more intricate in South Asian contexts, particularly in Bangladesh, where rapidly expanding internet access has opened new avenues for political participation even as economic status, literacy levels, religious values, and class distinctions significantly mediate voter behavior online; for example, while political parties such as the Awami League and the Bangladesh Nationalist Party (BNP) have effectively utilized social media to rally support, concerns over pervasive government surveillance under the Digital Security Act (DSA) have forced activists and journalists to self-censor, thereby constraining the free flow of information and reinforcing existing power structures (Human Rights Watch, 2021). Similar patterns are observed in neighboring countries like India and Pakistan, where digital platforms have been harnessed both to mobilize grassroots political energies and to propagate state-backed narratives, yet the interplay of communal tensions, religious identity, and hierarchical social structures further complicates the landscape by influencing how misinformation and hate speech circulate and ultimately affect electoral behavior. The pervasive threat of misinformation and manipulation exemplified by the proliferation of fake news, automated bots, paid trolls, and deepfakes poses a core challenge to the integrity of democratic processes worldwide, as these tools create an environment of mistrust and confusion that undermines confidence in legitimate news sources and fuels conspiratorial thinking, a phenomenon that is particularly acute in emerging democracies lacking robust media infrastructures and widespread digital literacy (Chesney & Citron, 2019). The synergy between platform algorithms optimized for engagement and polarized user communities amplifies these phenomena (D3), creating filter bubbles and echo chambers that facilitate the rapid spread of disinformation among ideologically homogeneous groups, making it extremely difficult to correct false narratives once they have become entrenched; although developed democracies like the United States have begun to implement fact-checking initiatives and content moderation strategies, these measures are often inconsistent and face resistance on free speech grounds, thereby highlighting the urgent need for more systematic interventions (Kosseff, 2019). In the realm of regulatory frameworks, the contrast between approaches in developed and emerging democracies is stark: in the United States, Section 230 of the Communications Decency Act has long provided platforms with immunity from liability for user-generated content, promoting a model of self-regulation that fosters innovation but also enables the unchecked spread of misinformation, while in Bangladesh, the DSA grants broad powers to state authorities to censor and prosecute online speech an approach that, although aimed at maintaining public order, has frequently been criticized for stifling dissent and curbing press freedom (Human Rights Watch, 2021). This tension between protecting civil liberties and curbing harmful content is further complicated by the rise of algorithmic governance, where opaque machine-learning systems are tasked with moderating, ranking, and recommending content; these systems, while essential for managing the vast amounts of data generated online, are susceptible to biases that can disproportionately amplify sensational or extremist content, thereby necessitating urgent calls for greater transparency, independent audits, and accountability measures (Pasquale, 2015). Against this backdrop, policy interventions aimed at strengthening democratic resilience and enhancing digital literacy have emerged as critical tools for mitigating the negative effects of digital disruptions on electoral processes; such interventions include comprehensive media literacy programs that integrate critical thinking and digital analysis skills into school and university curriculums, as well as community initiatives and NGO-led training sessions that target underserved populations, thereby equipping citizens with the ability to discern credible information from disinformation (Mihailidis & Viotty, 2017). In parallel, regulatory measures that focus on ensuring transparency in political advertising such as mandating clear disclosures on digital ads, imposing spending caps, and requiring comprehensive transparency reports are essential for demystifying the often opaque world of targeted political communication, a practice that has been linked to the manipulation of voter behavior through micro-targeted campaigns (Kreiss & McGregor, 2019). Moreover, strengthening civil society and independent media is paramount; robust watchdog journalism, fact-checking platforms, and citizen reporting initiatives not only hold governments and digital platforms accountable but also provide alternative sources of information that counterbalance the manipulative narratives often disseminated online (Reporters Without Borders, 2021). Finally, global partnerships and international coalitions epitomized by initiatives such as the EU Digital Services Act,

UNESCO's Internet Universality Indicators, and the Global Network Initiative offer promising avenues for sharing regulatory best practices and technical expertise across borders, fostering a transnational framework that addresses digital disinformation and algorithmic biases while promoting human rights and democratic values (European Commission, 2022; UNESCO, 2020). Taken together, these insights underscore that the transformation from traditional media to digital platforms has ushered in both opportunities and risks: while digital media can democratize political participation by empowering grassroots voices and facilitating diverse debates, it simultaneously poses significant challenges through the spread of misinformation, the exploitation of algorithmic systems, and the imposition of repressive regulatory measures that can undermine civic freedoms. The evolving digital polity is thus characterized by a series of paradoxes, in which the tools that promise greater inclusivity and accountability are the same ones that can be manipulated to entrench power, distort electoral processes, and silence dissent. Ultimately, the future of democratic engagement in the digital age will depend on our ability to implement multifaceted policy interventions that integrate technical innovation with ethical oversight, foster media literacy at every level of society, and promote international collaboration to harmonize regulatory practices efforts that are essential for building a resilient, transparent, and inclusive digital public sphere capable of withstanding the manifold challenges of the twenty-first century.

10.4 Reflecting on Digital Media's Paradoxical Role

One of the central paradoxes spotlighted throughout this book is the dual character of digital media. On the one hand, platforms like Facebook, Twitter (X), and WhatsApp democratize information flow, enabling grassroots mobilizations that challenge traditional gatekeepers (Tufekci, 2017). Citizen journalists and marginalized groups now have direct channels to share their perspectives, circumventing state-friendly mainstream outlets. This democratizing potential is evident in the success of movements like #MeToo or various protest campaigns in the Middle East, Europe, and Latin America, where social media galvanized large-scale collective action. Yet, this open environment also yields vulnerabilities. Misinformation travels faster online than corrections, fueling rumor cascades and conspiratorial thinking (Vosoughi et al., 2018). Voters can be micro-targeted with manipulative ads that distort public debate. Authoritarian regimes or illiberal actors exploit these same tools to surveil dissidents, distribute propaganda, and disrupt electoral processes. The U.S. experience with Russian interference in 2016, or Bangladesh's heavy-handed crackdown under the DSA, underscores how digital media can be subverted into a tool for controlling narratives. The tension between liberation and manipulation thus emerges as a defining feature of contemporary political life online. While digital technologies broaden access to information, they also produce fragmentation via filter bubbles and echo chambers (Pariser, 2011; Sunstein, 2017). Algorithmic personalization fosters insular communities around shared beliefs, reducing exposure to conflicting viewpoints. In polarized contexts, these enclaves intensify partisanship and entrench citizens in self-reinforcing belief systems. This phenomenon is especially pronounced in advanced democracies like the United States, where ideological splits align with consumption patterns on social platforms. Emerging democracies face a variant of this challenge: communal or religious lines shape digital enclaves, sometimes triggering violent offline consequences if hate speech or incendiary rumors circulate unchecked (Ahmed, 2020). At the same time, digital media can unify diverse groups around shared concerns, bypassing partisan mainstream outlets. Hashtags like #BlackLivesMatter or #RoadSafetyBD illustrate how viral activism can transcend traditional boundaries. The paradox is that while the same networked infrastructure can produce cohesive protest communities, it can also entrench divisions if manipulative actors inflame cultural cleavages for political gain. Thus, fostering rational discourse online demands strategies that acknowledge both the integrative and fragmenting tendencies of social media platforms. Another paradox is visible in debates over regulating extremist content, disinformation, and foreign electoral meddling. Governments claim a duty to safeguard citizens against malicious or destabilizing digital activity, yet their interventions often risk infringing on civil liberties. The U.S. approach, with its broad free speech protections under the First Amendment, historically favored minimal oversight, only to witness challenges from extremist recruitment, QAnon conspiracies, and alleged foreign interference (Chesney & Citron, 2019). Meanwhile, Bangladesh's top-down legislation, e.g., the Digital Security Act, is criticized for chilling legitimate dissent under the guise of public order (Human Rights Watch, 2021). Striking a balance requires nuanced frameworks that impose obligations on platforms to curtail genuinely harmful or illegal content while protecting lawful speech. It also entails preventing state abuse of censorship powers that suppress political opposition. Proposed solutions include co-regulation, independent oversight bodies, judicial safeguards, and user empowerment via encryption tools. However, no single model perfectly resolves tensions between security interests and the freedoms that define democratic societies. As digital platforms increasingly rely on automated systems to moderate vast amounts of user-generated content, algorithmic complexity introduces new dilemmas. Machine-learning algorithms can inadvertently reproduce biases present in training datasets, systematically marginalizing certain groups or perpetuating stereotypes (Noble, 2018). The opacity of such black-box models complicates accountability users, regulators, and even platform operators themselves may struggle to explain why certain content is promoted or removed. Solutions proposed in earlier chapters, such as algorithmic audits and transparency reports, remain experimental and face corporate resistance (Pasquale, 2015). Moreover, platforms must balance user privacy with moderation demands. Automated content scanning to detect hate speech or extremist threats can encroach on personal communications. The swirl of controversies over encryption "backdoors," mass surveillance, and content scanning exemplifies how algorithmic governance intersects with fundamental rights. Future research must clarify how accountability measures can function across different cultural and political contexts, ensuring that interventions do not unwittingly expand state or corporate surveillance.

10.5 Proposing Future Research Trajectories

Building on the insights gleaned from earlier chapters, several promising avenues for interdisciplinary research emerge that could substantially deepen our understanding of digital media's multifaceted impact on political behavior and governance. First, cross-platform comparative studies should move beyond analyses confined to individual sites like Facebook or Twitter and instead investigate how users navigate seamlessly between public spaces and private channels such as WhatsApp, Telegram, or Signal, thereby illuminating the complex ways content flows across environments and influences electoral behavior and protest dynamics (Chadwick, 2013). In tandem, algorithmic auditing methods require the development of replicable frameworks for conducting independent evaluations of recommendation engines, content moderation protocols, and ad-targeting algorithms; designing "audit sandboxes" where academic teams collaborate confidentially with platform engineers could offer critical insights into data patterns and biases (Pasquale, 2015). Moreover, while media literacy programs have shown early promise, little is known about their long-term impact once participants leave formal educational settings, so longitudinal studies tracking cohorts over multiple election cycles are needed to assess whether such interventions lead to sustained improvements in discerning fake news, resisting manipulative ads, and engaging constructively in political debate (Mihailidis & Viotty, 2017). At the same time, contextualized misinformation studies should examine how deeply entrenched cultural, religious, or ideological tensions shape the spread of rumors in diverse settings comparing rural versus urban, high-literacy versus low-literacy, and homogeneous versus multi-ethnic communities to understand the nuanced efficacy of various interventions, particularly in countries like Bangladesh, India, or Nigeria (Ahmed, 2020). Additionally, rigorous policy experimentation and impact evaluations employing quasi-experimental or experimental methods, such as randomized controlled trials, could test the real-world effectiveness of proposed interventions like disclaimers on political ads, spending caps, or co-regulatory oversight boards, thereby guiding policymakers in refining these strategies and identifying unintended consequences (European Commission, 2022). Furthermore, expanding research to include comparative analyses of authoritarian versus democratic regimes would illuminate how state control over the internet alters digital mobilization strategies, exploring how activists mobilize through alternative or encrypted channels and how state propaganda adapts to censorship-evading technologies, thus identifying structural variables that predict resilience or vulnerability to digital authoritarianism (Freedom House, 2022). Finally, fostering interdisciplinary collaboration is crucial; bringing together political scientists, communication experts, sociologists, computer scientists, legal scholars, and psychologists in collaborative labs can yield robust analytics that bridge large-scale computational methods with nuanced interpretive frameworks, ultimately producing policy-relevant insights that integrate human rights, ethical design, and political feasibility (Chesney & Citron, 2019).

10.6 Interdisciplinary Collaborations for Continued Study

Political scientists bring theories of electoral behavior, party competition, and institutional design, while communication scholars analyze agenda-setting, framing, and media effects. Joint projects could examine how micro-targeted campaigning influences voter turnout, partisan identification, or policy preferences. Comparative lenses across multiple countries would help clarify how varied regulatory backdrops or cultural norms mediate these phenomena (Sunstein, 2017). Sociologists and anthropologists offer insights into how digital networks intersect with social structures, identity formation, and cultural practices. Ethnographic fieldwork can reveal how rumor cascades unfold in localized settings, how religious or communal leaders sway digital discourse, or how youth subcultures adopt new platforms. In emerging democracies, sociologists might focus on how vulnerable communities navigate misinformation or censorship, illuminating grassroots coping mechanisms (Ahmed, 2020). Algorithmic complexity demands technical expertise to interpret the "black boxes" of content moderation, recommendation systems, and data analytics. Computer scientists can develop advanced tools for detecting coordinated inauthentic behavior, analyzing sentiment dynamics, or identifying deepfakes. Collaborations with domain experts in politics and media ensure that these tools address real-world concerns like discerning manipulative advertising or lethal hate speech while respecting user privacy and ethical constraints (Noble, 2018). Law and policy scholars scrutinize legislative proposals, court rulings, and regulatory enforcement. Collaboration with communication experts can clarify the speech implications of regulating platform content or imposing liability for misinformation. Meanwhile, policy analysts evaluate cost-effectiveness, feasibility, and stakeholder interests to propose balanced interventions that mitigate digital threats without stifling innovation or speech. Partnerships between academia, think tanks, and legislative bodies can accelerate the translation of research findings into actionable laws (Kosseff, 2019). Finally, human rights advocates and ethicists provide frameworks for balancing national security, corporate interests, and civil liberties. Digital technology raises ethical dilemmas regarding privacy, autonomy, and the potential for algorithmic discrimination. Collaborative efforts can produce guidelines or certification programs that encourage ethical design principles, data minimization, and robust user consent. In contexts like Bangladesh, where the state can exercise sweeping digital surveillance, ethical frameworks become crucial for safeguarding basic freedoms (Human Rights Watch, 2021; UNESCO, 2020).

10.7 Notes for Instructors and Researchers

The insights offered across these chapters are suitable for graduate-level courses in Political Science, Communication, Sociology, and International Development. Each chapter weaves together theoretical underpinnings and empirical case studies, making it adaptable to seminar-style discussions, case analyses, or capstone projects. Instructors might assign individual chapters to explore distinct but interconnected topics: historical media evolution, the interplay of media and democracy in South Asia, misinformation challenges, policy interventions, and more. One pedagogical approach is to invite students to conduct small-scale "digital fieldwork": analyzing local social media groups or mapping how a political controversy unfolded online. For instance, a course might culminate in group projects that develop policy proposals or draft digital literacy curricula for real or hypothetical communities. Such activities promote critical thinking, encourage active learning, and allow students to grasp the complexities of digital governance. Beyond classroom applications, these chapters lay a foundation for empirical data gathering and comparative analysis. Researchers might employ surveys, focus groups, social media analytics, or natural experiments to deepen our understanding of how digital media shapes voter behavior in Bangladesh, the U.S., and beyond. Multi-country studies could expand on the frameworks introduced, comparing, for example, how India, Pakistan, or Nigeria navigate digital disruptions within distinctive cultural, economic, and political contexts. Scholars might also investigate longitudinal effects of media literacy interventions, or evaluate how different regulatory environments like the U.S. self-regulatory tradition vs. the EU's co-regulatory model affect rates of misinformation, online harassment, or voter confidence. Cross-national collaboration, facilitated by networks such as the International Communication Association or the International Political Science Association, can yield robust datasets that transcend individual case peculiarities. From a policy standpoint, the concluding chapters underscore concrete interventions to reconcile digital freedoms with the imperative to safeguard electoral integrity. Recommendations such as mandatory disclosures on political ads, algorithmic transparency, and establishing independent oversight boards speak directly to ongoing legislative debates worldwide (European Commission, 2022). Policymakers grappling with issues of censorship, surveillance, or foreign electoral interference may draw inspiration from the multi-tiered approaches outlined here, which balance civic empowerment, accountability mechanisms, and respect for civil liberties. Strengthening institutions remains paramount. Election commissions can partner with digital platforms to monitor campaign spending, while judicial bodies can refine legal standards that limit the overreach of anti-cybercrime laws. Cross-border dialogues perhaps via regional bodies like ASEAN or the African Union could harmonize policies to tackle transnational networks of disinformation. Meanwhile, non-state actors such as NGOs and philanthropic institutions are well-positioned to fund digital literacy initiatives, fact-checking collaborations, and research that illuminates evolving digital threats.

10.8 The Road Ahead: Unresolved Questions and Emerging Challenges

Building on the insights from earlier chapters, the road ahead in digital governance is fraught with unresolved questions and emerging challenges that will shape the evolving digital polity, demanding ongoing interdisciplinary collaboration and iterative policy refinement. Technological advances in AI-based media generation particularly the proliferation of deepfakes and synthetic media continue to lower the barriers to producing hyper-realistic audio and video fabrications, thereby escalating the potential for sowing confusion, defaming political figures, and undermining public trust; these developments raise urgent questions about the adequacy of current detection methods, the availability of effective legal recourse, and the responsibilities that platforms must assume in mitigating such risks (Chesney & Citron, 2019). At the same time, governments worldwide remain deeply divided over the role of encryption: while robust encryption is essential for protecting civil liberties and ensuring secure communications, many state actors demand "exceptional access" to encrypted data for law enforcement and national security purposes a tension that is particularly pronounced in emerging democracies, where stringent measures to weaken encryption may lead to unchecked surveillance and further erode citizen privacy (EFF, 2021). Moreover, the issue of data governance and privacy has become increasingly critical as personal data underpins targeted political advertising and micro-targeting strategies; although regulatory frameworks such as the EU's General Data Protection Regulation (GDPR) and California's Consumer Privacy Act have set important precedents, many jurisdictions still lack comprehensive legislation, leaving policymakers with the complex task of balancing user privacy, sustaining viable business models, and ensuring political transparency (Vosoughi et al., 2018). Adding to these concerns is the immense power held by digital platforms, which act as de facto gatekeepers by curating and prioritizing content through opaque algorithmic processes; while self-regulation enables rapid adaptation to emerging issues, corporate profit motives often constrain meaningful accountability, and attempts by governments to impose external oversight risk stifling innovation or curtailing free speech, thus presenting a core policy dilemma of the digital era (Kosseff, 2019). These challenges underscore that the digital polity is far from static it evolves in tandem with technological progress and shifting political landscapes, necessitating an ongoing commitment to rigorous research and global cooperation. International organizations such as UNESCO and Reporters Without Borders, alongside multi-stakeholder initiatives like the Global Network Initiative (GNI), play a pivotal role in establishing transnational norms and best practices by fostering dialogue, capacity building, and the exchange of technical expertise, which are essential for harmonizing regulatory responses across diverse political contexts (UNESCO, 2020). In light of these complexities, future research must explore several critical avenues, including the development of replicable algorithmic auditing methods, the longitudinal impact of digital literacy programs on voter behavior, and contextualized studies that examine how cultural, religious, and socioeconomic factors mediate the spread of misinformation across different regions such as in Bangladesh, India, or Nigeria (Ahmed, 2020; Mihailidis & Viotty, 2017). Ultimately, the convergence of these challenges calls for a hybrid regulatory approach that blends robust government oversight with market-driven self-regulation, ensuring that transparency measures like mandated disclosures for political ads, spending caps, and independent oversight of algorithmic processes are implemented without compromising the core democratic values of free speech and innovation. As digital media continues to reshape public discourse and electoral processes, the future of digital governance will depend on our collective ability to balance competing imperatives, protect civil liberties, and foster a resilient, inclusive digital public sphere that supports both democratic participation and societal cohesion.

10.9 Toward a More Equitable and Resilient Digital Polity

To ensure that digital media fulfills its democratic promise without succumbing to the pitfalls of disinformation, manipulation, and polarization, societies must adopt multi-level strategies that integrate technical innovation with ethical and political oversight, engaging educators, policymakers, civil society, the private sector, and international organizations in a concerted effort to create a resilient digital public sphere. Educators can play a transformative role by mainstreaming digital literacy from an early age encouraging critical thinking, media analysis, and civic responsibility which is crucial for enabling citizens to navigate complex online environments and discern credible information from misleading content (Mihailidis & Viotty, 2017). At the same time, policymakers are called upon to institute transparent rules governing political advertising, to champion robust data protection laws, and to establish oversight bodies that include citizen representation, ensuring that regulatory frameworks do not merely serve the interests of powerful actors but protect public trust and electoral integrity (Kreiss & McGregor, 2019). Civil society organizations, meanwhile, must amplify watchdog functions, foster inclusive debates, and actively challenge instances of corporate malfeasance and state overreach, thereby acting as essential counterbalances to potential abuses in the digital arena (Reporters Without Borders, 2021). Private sector actors, especially major digital platforms, should commit to ethical design principles by embracing independent algorithmic audits and adopting consistent community standards that guard against manipulation and bias, ensuring that their technologies serve public interests rather than exacerbating societal divisions (Pasquale, 2015). International organizations, such as UNESCO and the Global Network Initiative, can unify best practices, coordinate cross-border strategies against online extremism, and build capacity in regions where local resources are limited, thereby fostering global standards that support transparency and human rights in digital communications (UNESCO, 2020). Collectively, these coordinated efforts can help realize a digital polity that values openness, inclusivity, and resilience an imperative in an era where complex challenges such as climate change, pandemics, and global inequality demand coordinated international responses, making the integrity and transparency of digital communication critical to forging consensus and sustaining democratic governance worldwide.

10.10 A Final Reflection: Embracing the Ambiguities of the Digital Age

Throughout this book, we have observed that digital media's capacity to empower citizens and catalyze political transformation is inextricably linked to its potential to disrupt, polarize, and enable authoritarian tendencies. This duality need not lead to despair; rather, it challenges us to craft nuanced, context-sensitive strategies that harness digital tools for constructive engagement while mitigating harm. The tension between free expression and content moderation, between open markets and regulatory safeguards, and between sovereignty and transnational digital networks reflects deeper questions about how 21st-century societies organize power and accountability. Embracing this ambiguity requires an openness to experimentation and learning from both successes and failures. The U.S. experience with minimal government intervention demonstrates the perils of unchecked corporate power over public discourse, while Bangladesh's top-down approach highlights the risks of censorship under the pretext of stability. Neither extreme offers a complete solution, suggesting that a balanced approach co-regulation or multi-stakeholder governance holds the greatest promise. In closing, we must acknowledge that technology alone does not determine political outcomes; human agency and institutional frameworks remain decisive. Even in a hyper-connected age, democratic values like transparency, equality, and justice guide how digital infrastructures evolve and how social actors negotiate their usage. The evolving digital polity is thus as much a reflection of collective societal choices as it is a product of technological capabilities. By continuing to ask critical questions, maintain scholarly rigor, and innovate collaboratively, researchers, practitioners, and citizens can steer digital media toward strengthening, rather than undermining, democratic life.

10.11 Concluding Call: Sustaining Dialogue and Research

The chapters in this volume collectively underscore the significance of digital media in shaping elections, governance, and civic discourse. Yet, they also highlight the complexities, trade-offs, and ethical dilemmas that arise when technology intersects with political power. To navigate these challenges, we encourage readers whether graduate students, scholars, policymakers, or engaged citizens to remain active participants in this evolving conversation. Rigorous research, comparative studies, and policy experimentation will be essential to refining our strategies for a more equitable and resilient digital future. Hence, the concluding call is this: continue the dialogue. Integrate these insights into academic curricula that blend theory with practical analysis, fuel new lines of empirical inquiry, and champion reforms that protect the open exchange of ideas. Collaboration between academia, government, and civil society can yield robust, evidence-driven policies that empower the many rather than the few. With sustained attention, creativity, and cross-disciplinary effort, we can shape a digital polity that upholds democratic ideals and paves the way for inclusive, transparent, and responsible political engagement in the decades to come.

10.12 Guidance for Instructors and Researchers: Integration, Research Extensions, and Policy Relevance

These concluding notes for instructors and researchers emphasize that the synthesis provided throughout this research book is exceptionally well-suited for integration into graduate-level curricula in Political Science, Communication, Sociology, and International Development, offering comprehensive theoretical discussions, detailed case study analyses, policy-oriented debates, and empirical references that can anchor both lectures and seminar discussions. The book's theoretical underpinnings ranging from Habermas' public sphere and agenda-setting theory to the concept of filter bubbles offer foundational insights into how media technology intersects with democracy, inviting students to critically debate the relevance of these frameworks amid radical digital transformations. Empirical examples drawn from the U.S. and Bangladesh serve as comparative anchors, with the potential for instructors to incorporate or assign cross-national case studies from countries such as India, Pakistan, Brazil, or Nigeria, thereby broadening the scope of classroom discussions through the inclusion of diverse cultural and institutional contexts. Instructors may also enrich the learning experience by incorporating guest speakers, role-playing simulations, and interactive digital literacy workshops. For research extensions, there is significant potential for empirical data gathering through surveys and focus groups that investigate voter perceptions of social media ads, misinformation, or censorship in specific electoral contexts, while social media analytics can be employed using platform APIs or third-party tools to map political communication networks, identify trending hashtags, or detect suspicious bot activity in real time. Longitudinal studies are also encouraged to track changes in political engagement or opinions across multiple election cycles, correlating these shifts with media literacy interventions or the implementation of new regulatory measures. Moreover, comparative frameworks could extend the analysis beyond Bangladesh and the U.S. to include other rapidly digitizing democracies such as Indonesia, Kenya, or Mexico, and by contrasting authoritarian regimes with open societies, researchers can uncover structural variables that either accelerate or restrain digital propaganda campaigns. From a policy relevance perspective, scholars are urged to examine the balance between freedoms and regulatory safeguards by evaluating how partial reforms like the EU's

Digital Services Act affect user privacy, transparency in political ads, and corporate compliance. Concrete interventions, such as mandating disclaimers on targeted political ads, requiring transparency reports, and enforcing algorithmic audits, are highlighted as vital for balancing freedom with accountability. Additionally, strategies aimed at strengthening institutions like enabling election commissions to track digital campaign expenditures or empowering judicial systems to adjudicate digital rights disputes are essential for sustaining fair competition, while cross-border dialogue platforms can facilitate the exchange of disinformation response tactics, best practices for algorithmic transparency, and insights into the role of diaspora networks in transnational politics, thereby driving more coordinated regulatory approaches globally.

Case Study

The book's comprehensive analysis reveals a paradox: digital media simultaneously empowers democratic participation and enables authoritarian control, facilitates grassroots mobilization and corporate manipulation, creates unprecedented information access and weaponized misinformation. The 2024 Bangladesh uprising toppled an authoritarian government through digital coordination, yet the Digital Security Act had previously silenced countless dissidents. Cambridge Analytica exploited Facebook for manipulation, yet the same platform enables marginalized communities to organize for justice.

This paradox cannot be resolved through technology alone—neither optimistic digital utopianism nor pessimistic technological determinism adequately captures digital media's fundamentally political nature. The digital polity's future depends on choices made by individuals, institutions, and societies: whether to prioritize surveillance or privacy, engagement or accuracy, innovation or accountability, efficiency or equity.

Questions for Discussion

1. How can democratic theory evolve to address political communication that operates through opaque algorithms rather than transparent public discourse?
2. What institutional reforms would strengthen democratic resilience against digital manipulation while preserving open communication?
3. Should access to digital communication infrastructure be considered a fundamental democratic right requiring universal provision?
4. How might future technological developments (AI, virtual reality, brain-computer interfaces) further disrupt democratic processes?
5. What role should interdisciplinary scholarship play in shaping technology governance that serves democratic values?

Food for Thought

If democracy requires informed deliberation, but digital ecosystems reward emotional intensity over reasoned debate, must we redesign either technology or democracy—and which is more feasible?

What We Learn

The digital polity remains contested terrain where democracy's future will be determined not by technology's inherent properties but by human choices about its design, regulation, and use. Sustainable solutions require integrating technical, political, ethical, and social dimensions—no single discipline holds all answers. The path forward demands humility about uncertainty, commitment to evidence-based policymaking, inclusion of diverse voices in governance decisions, and recognition that perfect solutions are impossible but continuous improvement is essential. Democracy in the digital age is not a destination but an ongoing process of negotiation, adaptation, and renewal.

References

Ahmed, S. (2020). Freedom of the press in Bangladesh: Challenges and prospects. *Journal of South Asian Studies, 10*(2), 75–98.
Chadwick, A. (2013). *The hybrid media system: Politics and power*. Oxford University Press.
Chesney, R., & Citron, D. K. (2019). Deepfakes and the new disinformation war: The coming age of post-truth geopolitics. *Foreign Affairs, 98*(1), 147–155.
EFF (Electronic Frontier Foundation). (2021). EFF's work on encryption. https://www.eff.org
European Commission. (2022). The Digital Services Act package. https://ec.europa.eu/digital-services-act
Freedom House. (2022). Freedom on the Net 2022. https://freedomhouse.org
Human Rights Watch. (2021). Muzzling dissent: The impact of Bangladesh's digital security act. https://www.hrw.org
Kosseff, J. (2019). *The twenty-six words that created the internet*. Cornell University Press.
Kreiss, D., & McGregor, S. C. (2019). Technology firms shape political communication. *Political Communication, 36*(3), 264–266.
Mihailidis, P., & Viotty, S. (2017). Spreadable spectacle in digital culture: Civic expression, fake news, and the role of media literacy. *American Behavioral Scientist, 61*(4), 441–454.
Noble, S. U. (2018). *Algorithms of oppression: How search engines reinforce racism*. NYU Press.
Pariser, E. (2011). *The filter bubble: What the internet is hiding from you*. Penguin Books.
Pasquale, F. (2015). *The black box society: The secret algorithms that control money and information*. Harvard University Press.
Reporters Without Borders. (2021). World Press freedom index: Bangladesh. https://rsf.org
Sunstein, C. R. (2017). *#Republic: Divided democracy in the age of social media*. Princeton University Press.
Tufekci, Z. (2017). *Twitter and tear gas: The power and fragility of networked protest*. Yale University Press.
UNESCO. (2020). Internet universality indicators: A framework for assessing internet development. https://unesdoc.unesco.org
Vosoughi, S., Roy, D., & Aral, S. (2018). The spread of true and false news online. *Science, 359*(6380), 1146–1151.

Correction to: Beyond Ballot Boxes

Correction to:
B. Ahmed and R. J. Samir, *Beyond Ballot Boxes*,
https://doi.org/10.1007/978-3-032-27350-5

In the original version of chapters 1 and 6, some citations were updated. The correction chapters has been updated with the changes.

The updated versions of these chapters can be found at
https://doi.org/10.1007/978-3-032-27350-5_1
https://doi.org/10.1007/978-3-032-27350-5_6

B. Ahmed and R. J. Samir, *Beyond Ballot Boxes*,
https://doi.org/10.1007/978-3-032-27350-5_11

MIX
Papier aus verantwortungsvollen Quellen
Paper from responsible sources
FSC® C105338

If you have any concerns about our products,
you can contact us on
ProductSafety@springernature.com

In case Publisher is established outside the EU,
the EU authorized representative is:
Springer Nature Customer Service Center GmbH
Europaplatz 3, 69115 Heidelberg, Germany

Printed by Libri Plureos GmbH
in Hamburg, Germany